Biotech's

DICTIONARY *Of* Toxicology

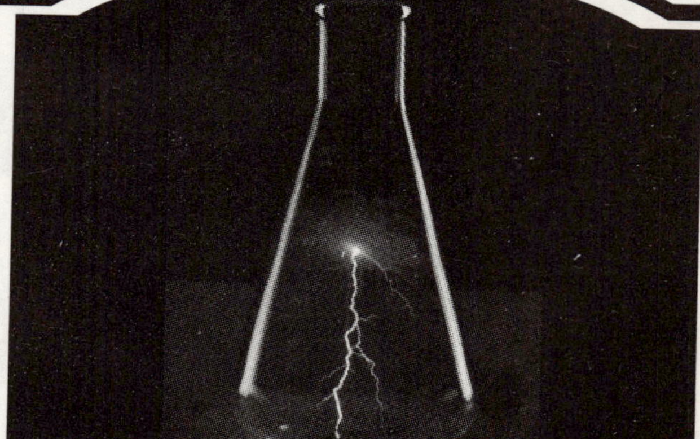

Compiled & Edited by

Dinesh Arora

BIOTECH

2004

BIOTECH BOOKS

DELHI - 110035

Biotech's

DICTIONARY

Of

Toxicology

Compiled & Edited by
Dinesh Arora

First Indian Edition 2004

ISBN 81-7622-118-X

Published by : **BIOTECH BOOKS**
1123/74, Tri Nagar,
Delhi - 110 035
Phone: 27382765
e-mail: biotechbooks@yahoo.co.in

Showroom : 4762-63/23, Ansari Road, Darya Ganj,
New Delhi - 110 002
Phone: 23245578, 23244987

Printed at : **Chawla Offset Printers**
New Delhi - 110 052

PRINTED IN INDIA

PREFACE

Toxicology, the study of poisons, is not a new subject. Spies and courtesans have used poison in their professional work for thousands of years and properties of different chemicals and their effects on human being has been a subject of interest to many.

However, modern toxicology is something more than disjointed information about individual venoms. It is a systematic science wide in its scope, covering ecology, industrial pollution and its effects on plants and animals as well as humans.

Green movement is an important radical movement of our times. Ecological consciousness is a defining element of our age. A desire to conserve our natural heritage and a lunatic desire to escape from industrial world have combined to produce this consciousness. As a scientific discipline, toxicology has much to offer to this movement and this age. A calm, considered evaluation of our options is the only way to move forward because wheels of history cannot be rolled back.

This dictionary covers a wide range of subjects from chemicals, ecology to cytology (study of cells), and physiology. A book to cherish and dip into from time to time.

- **abiological**
  see **abiotic.**
  'glue sniffing'
  · solvent abuse using plastic cement or other solvent-based adhesives. See **'solvent sniffing', addiction, dependence, solvent abuse.**
  'solvent sniffing'
  %t3 uptake
  a measurement of the unsaturated thyroxine binding sites of the serum thyroxine binding proteins.
  2,3-dimercaptopropan-1-ol
  metal chelator which has been used in the treatment of arsenic, antimony, gold, mercury and lead poisoning.
- **abiotic**
  this is a term used to describe anything which is characterised by the absence of life or incompatible with life. In toxicology and ecotoxicology it refers to physical (e.g., heat, sunlight) or chemical processes (e.g., hydrolysis) that are capable of modifying chemical structures. Thus, abiotic transformation is a process in which a chemical in the environment is modified by non-biological mechanisms.
- **abiotic transformation**
  process in which a substance in the environment is modified by non-biological mechanisms. See **biotransformation.**
- **absolute lethal concentration (LC_{100})**
  lowest concentration of a substance in an environmental medium which kills 100 % of test organisms or species under defined conditions. This value is dependent on the number of organisms used in its assessment.
- **absolute lethal dose (LD_{100})**
  lowest amount of a substance which kills 100 % of test animals under defined conditions. This value is dependent on the number of organisms used in its assessment.
- **absolute risk**
  the excess risk due to exposure to a hazard is referred to as the absolute risk.
- **absorbed dose**
  the actual amount of the exposed dose that enters the body.

- **absorbed dose (of radiation)**
  energy imparted to matter in a suitably small element of volume by ionising radiation divided by the mass of that element of volume. The SI unit for absorbed dose is joule per kilogram (Jkg^{-1}) and its special name is grey (Gy). See **ionising radiation.**

- **absorption**
  1. the process of taking in. For a person or an animal, absorption is the process of a substance getting into the body through the eyes, skin, stomach, intestines, or lungs.

Absorption Prediction System

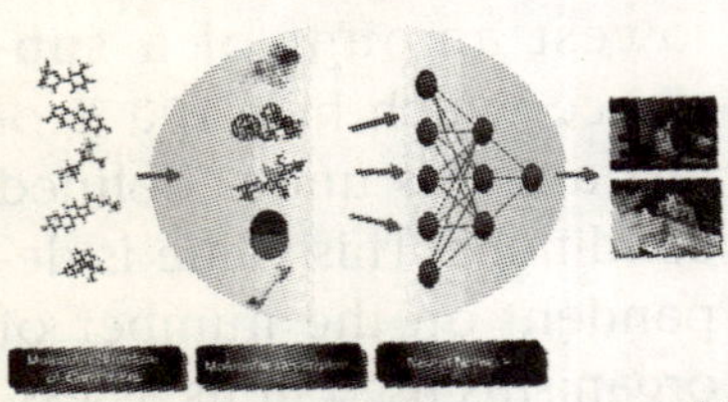

  2. the uptake of water or dissolved chemicals by a cell or an organism (as tree roots absorb dissolved nutrients in the soil).

- **absorption (biological)**
  process of active or passive transport of a substance into an organism: in the case of a mammal or human being, this is usually through the lungs, gastrointestinal tract, or skin.

- **absorption (in colloid and surface chemistry)**
  process whereby, when two phases are brought into contact, a particular component is transferred from one phase to the other.

- **absorption (of radiation)**
  phenomenon in which radiation transfers some or all of its energy to matter which it traverses.

- **absorption coefficient (in biology)**
  ratio of the absorbed amount (uptake)of a substance to the administered amount (intake): for exposure by way of the respiratory tract, the coefficient is the ratio of the absorbed amount to the amount of the substance (usually particles) deposited (adsorbed) in the lungs. See **absorbed dose, absorption factor.**

- **abuse (of drugs, substances, solvents etc.)**
  improper use of drugs or other substances. See **'glue sniffing', solvent abuse, 'solvent sniffing'**

- **acaricide**
  substance intended to kill mites, ticks or other Acaridae.

- **Acceptable Daily Intake (ADI)**
  estimate of the amount of a substance in food or drinking water, expressed on a body mass basis (usually mg/kg body weight), which can be ingested daily over a lifetime by humans without appreciable health risk. For calculation of the daily intake per person, a standard body mass of 60 kg is used. ADI is normally used for food additives (tolerable daily intake is used for contaminants). See **tolerable daily intake.**

- **acceptable residue level of an antibiotic**
  acceptable concentration of a residue which has been established for an antibiotic found in human or animal foods.

- **acceptable risk**
  probability of suffering disease or injury which is considered to be sufficiently small to be 'negligible'. See **tolerable risk, accepted risk, negligible risk, risk de minimis.**

- **accepted risk**
  probability of suffering disease or injury which is accepted by an individual.

- **accidental exposure**
  unintended contact with a substance or change in the physical environment (including for example radiation) resulting from an accident.

- **acclimation**
  1. the physiological and behavioural adjustments of an organism to changes in the environment.

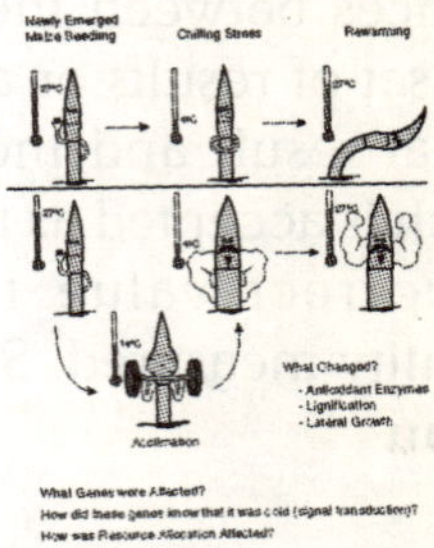

- **acclimatisation**
  1. the adaptation over several generations of a species to a marked change in the environment.
  2. processes, including selection and adaptation, by which a population of mi-

cro-organisms develops the ability to degrade a substance, or develops a tolerance to it.

3. in animal tests - allowing an animal to adjust to its environment prior to undertaking a study.

- **accumulation**
  successive additions of a substance to a target organism, or organ, or to part of the environment, resulting in an increasing amount or concentration of the substance in the organism, organ, or environment.

- **accuracy**
  quantity referring to the differences between the mean of a set of results or an individual result and the value which is accepted as the true or correct value for the quantity measured. See **precision**.

- **ACD**
  a solution of acid, citrate and dextrose used as an anti-coagulant and preservative in storing blood for transfusions.

- **acetaminophen**
  a pain reliever and fever reducer, the active ingredient in many medications. An overdose may be toxic.

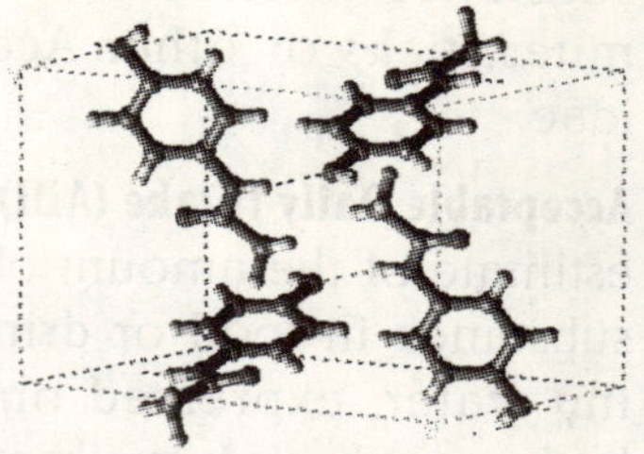

- **acetic acid**
  the acid found in vinegar, often used as a reagent. Also used in the preparation of pharmaceuticals, and in industry.

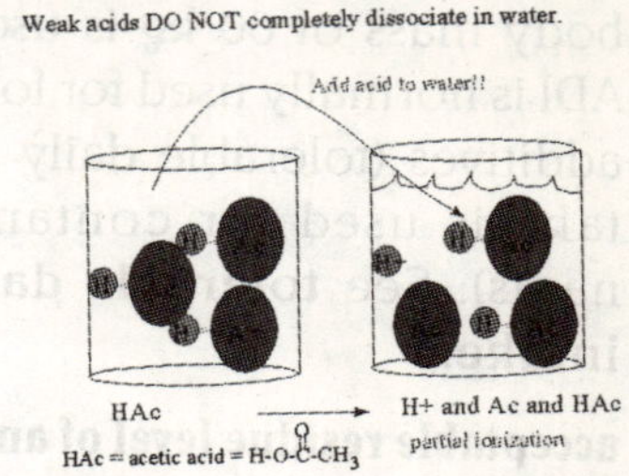

- **acetylcholinesterase (ACHE)**
  an enzyme present in various tissues, including muscle and red cells, that breaks down acetylcholine (a chemical released by nerves that activates muscle contractions) and helps to maintain proper transmission of impulses between nerve cells and between nerve cells and muscles, also called true cholinesterase. Measuring ace-

tylcholinesterase in amniotic fluid may help confirm a suspected neural tube defect in the foetus.

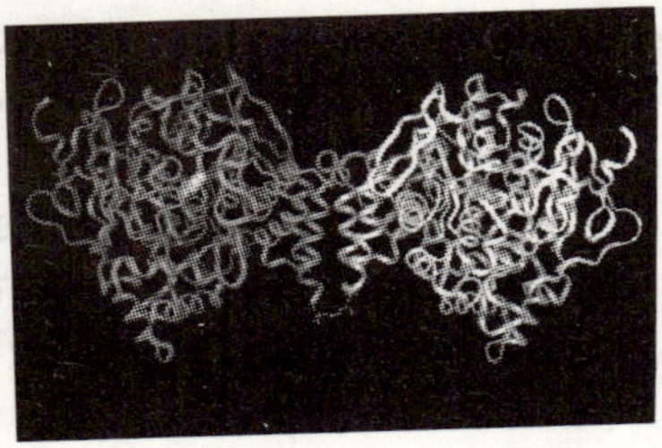

- **acetylsalicylic acid (ASA)**
  aspirin, salicylate. A pain reliever and fever reducer. An overdose can be toxic.

- **acid**
  a corrosive solution with a pH less than 7. Vinegar is a common weak acid. Battery acid is much stronger. An inorganic or organic compound that (1) reacts with metals to yield hydrogen (2) reacts with a base to form a salt (3) dissociates in water to yield hydrogen or hydronium ions (4) has a pH of less than 7.0 and (5) neutralises bases or alkaline media. All acids contain hydrogen and turn litmus paper red. They are corrosive to human tissue and are to be handled with care.

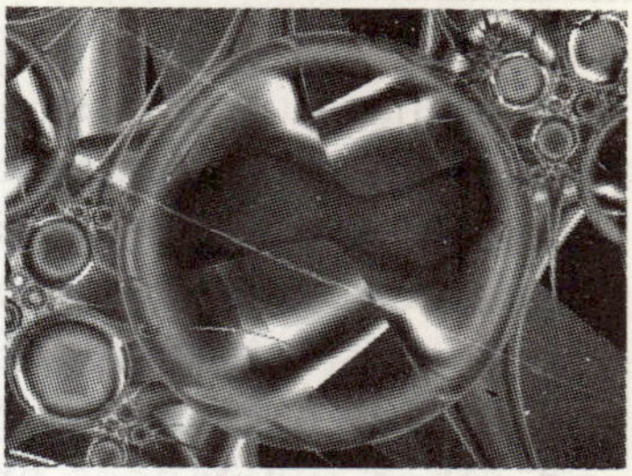

- **acid mine drainage**
  drainage of water from areas that have been mined for coal or other mineral ores: the water has low pH, sometimes less than 2.0 (is acid), because of its contact with sulphur-bearing material. Acid drainage is harmful because it often kills aquatic organisms.

- **acid phosphatase (acid phos)**
  an enzyme found primarily in the prostate and semen. Increased blood serum levels may indicate cancer of the prostate or may follow prostatic massage.

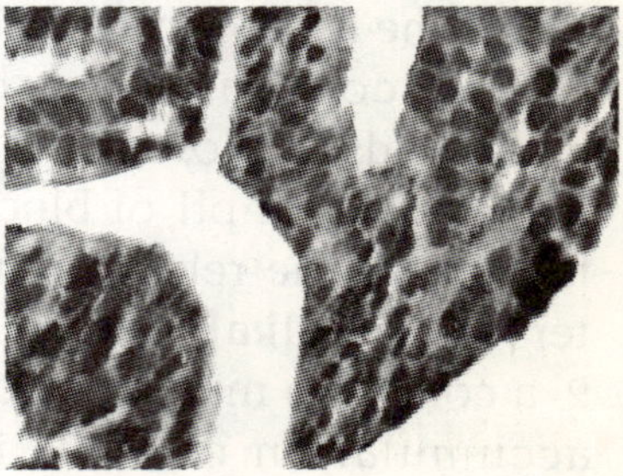

- **acid rain**
you first need to understand Acid Deposition: a complex chemical and atmospheric phenomenon that occurs when emissions of sulphur and nitrogen compounds and other substances are transformed by chemical sources, and then deposited on earth in either wet or dry form. The wet forms (precipitation) are popularly called 'acid rain' and fall as rain, snow, or fog. The dry forms are acidic gases or particulates.

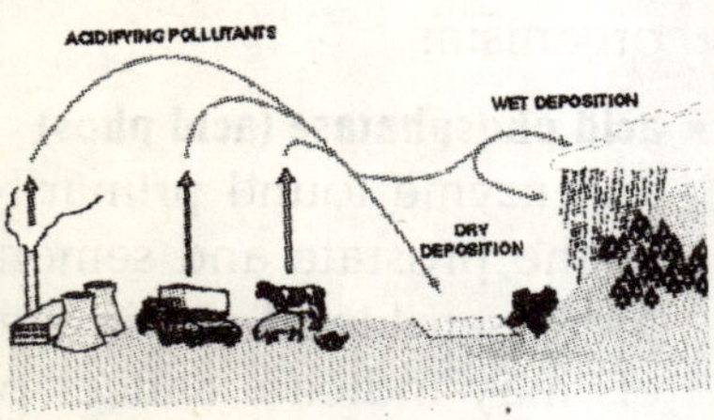

- **acidosis**
1. pathological condition in which the hydrogen ion substance concentration of body fluids is above normal and hence the pH of blood falls below the reference interval. See **alkalosis**.
2. a condition marked by an accumulation of acids in body tissues and blood that may result from prolonged diarrhoea or disorders such as diabetes and kidney disease, when not under control. Urine tests screen for acidosis. Determination of blood pH provides more accuracy.
3. a condition of decreased alkalinity of the blood and tissues marked by sickly sweet breath, headache, nausea, vomiting, and visual disturbances, usually the result of excessive acid production. Tissues and CNS functions are disturbed.

- **Acquired Immunodeficiency Syndrome**
see **AIDS**.

- **action level**
1. concentration of a substance in air, soil, water or other defined medium at which specified emergency counter-measures, such as the seizure and destruction of contaminated materials, evacuation of the local population or closing down the sources of pollution, are to be taken.
2. concentration of a pollutant in air, soil, water or other defined medium at which some kind of preven-

tive action (not necessarily of an emergency nature) is to be taken.

- **activated carbon**

adsorptive particles or granules of carbon usually obtained by heating carbon (such as wood). These particles or granules have a high capacity to selectively remove certain trace and soluble materials from water

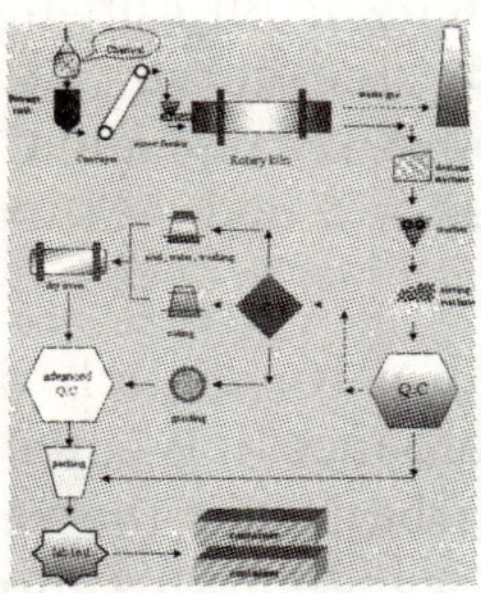

Activated Carbon Processing

- **activated sludge process**

a sewage treatment process by which bacteria that feed on organic wastes are continuously circulated and put in contact with organic waste in the presence of oxygen to increase the rate of decomposition.

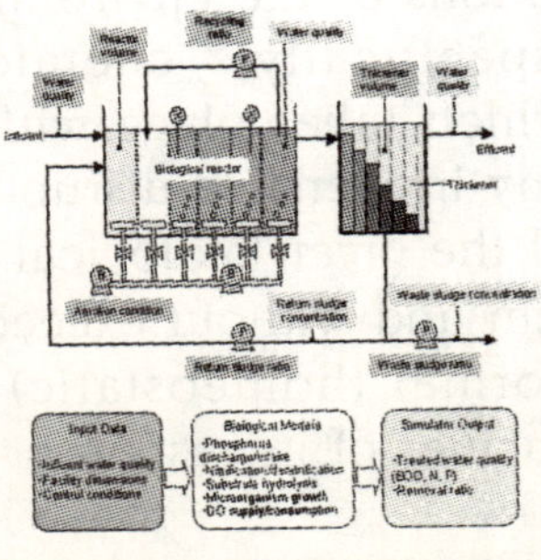

- **activation**

see **bioactivation**.

- **active ingredient**

in any pesticide product, the component that kills, or otherwise controls, target pests. Pesticides are regulated primarily on the basis of active ingredients.

- **acute**

short-term, in relation to exposure or effect. In experimental toxicology, 'acute' refers to studies of two weeks or less in duration (often less than 24 h).

- **acufe effect**

effect of short duration and occurring rapidly (usually in the first 24 h or up to 14 d) following a single dose or short exposure to a substance or radiation.

- **acute exposure**

contact with a substance that occurs once or for only a short time (up to 14 days).

- **acute test**

a comparative study in which organisms, that are

subjected to different treatments, are observed for a short period usually not constituting a substantial portion of their life span. Acute tests often utilise mortality as the only measured effect. Chronic tests usually include additional measures of effect such as growth or reproduction .

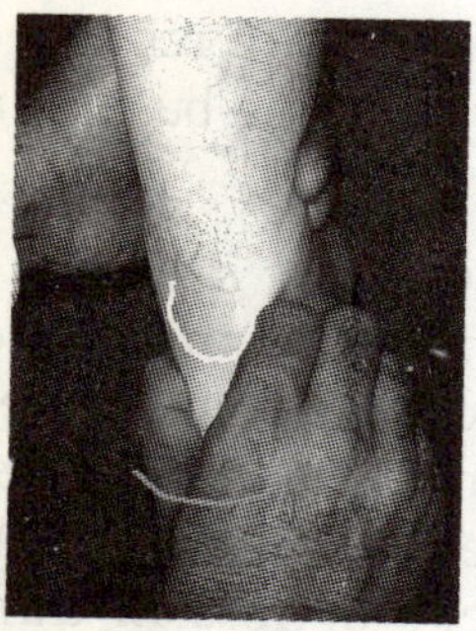

- **acute toxicity**
  adverse effects occurring within a short time (usually up to 14 d) after administration of a single dose (or exposure to a given concentration)of a test substance or after multiple doses (exposures), usually within 24 h.
  Ability of a substance to cause adverse effects within a short time of dosing or exposure. See **chronic toxicity**.

- **acute toxicity test**
  experimental animal study to determine what adverse effects occur in a short time (usually up to 14 d) after a single dose of a substance or after multiple doses given in up to 24h.

- **acute-to-chronic ratio (ACR)**
  the ratio of the acute toxicity to chronic toxicity of a chemical or sample which can be used to predict acute toxicity from chronic data and vice-versa. Example LC50 = 100 ug/l and Chronic Value = 5 ug/l, ACR = 20.

- **AD**
  symbol for a dichromatic absorbance. It is equal to the absorbance at the secondary wavelength subtracted from the absorbance at the primary wavelength.

- **adaptation**
  1. change in an organism, in response to changing conditions of the environment (specifically chemical), which takes place without any irreversible disruptions of the given biological system and without exceeding normal (homeostatic) capacities of its response.

2. process by which an organism stabilises its physiological condition after an environmental change. See **acclimatisation**.

- **added risk**
  difference between the incidence of an adverse effect in a treated group (of organisms or a group of exposed humans) and a control group (of the same organisms or the spontaneous incidence in humans).

- **addiction**
  surrender and devotion to the regular use of a medicinal or pleasurable substance for the sake of relief, comfort, stimulation, or exhilaration which it affords, often with craving when the drug is absent. See **dependence**.

- **additive effect**
  consequence which follows exposure to two or more physico-chemical agents which act jointly but do not interact, commonly, the total effect is the simple sum of the effects of separate exposure to the agents under the same conditions. Substances of simple similar action may show dose or concentration addition.

- **additivity**
  the response to a combination of two or more chemicals is the sum of the expected individual responses

- **adduct**
  new chemical species AB, each molecular entity of which is formed by direct combination of two separate molecular entities A and B in such a way that there is no change in connectivity of atoms within their moieties A and B. Stoichiometries other than 1:1 are also possible. An intramolecular adduct can be formed when A and B are groups contained within the same molecular entity.

- **adenocarcinoma**
  malignant tumour originating in glandular epithelium

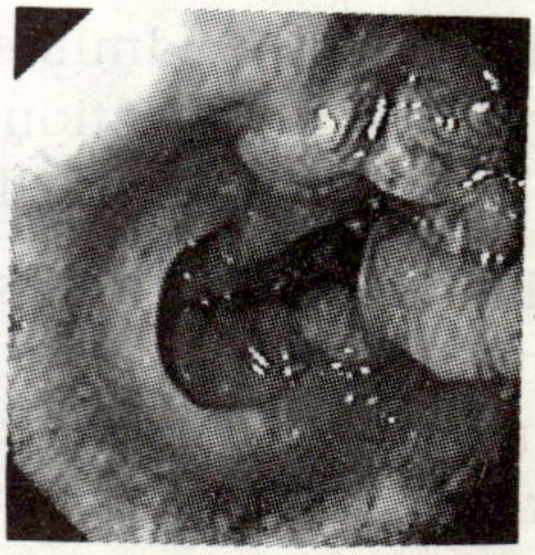

or forming recognisable glandular structures. See **adenoma**.

- **adenoma**
benign tumour occurring in glandular epithelium or forming recognizable glandular structures.

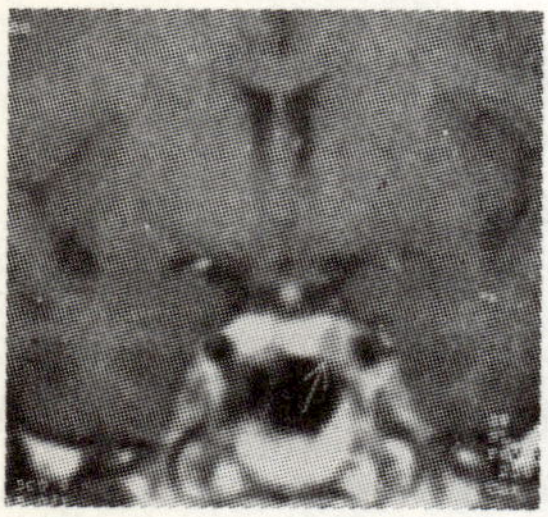

- **adjuvant**
1. in pharmacology, a substance added to a drug to speed or increase the action of the main component.
2. in immunology, a substance (such as aluminium hydroxide) or an organism (such as bovine tuberculosis bacillus) which increases the response to an antigen.

- **administered dose**
the quantity administered orally or by injection (usually in testing or medicine).

- **administration (of a substance)**
application of a known amount of a substance to an organism in a reproducible manner and by a defined route.

- **adrenal glands**
located near the kidneys, the adrenal glands produce a small amount of the male hormone, testosterone.

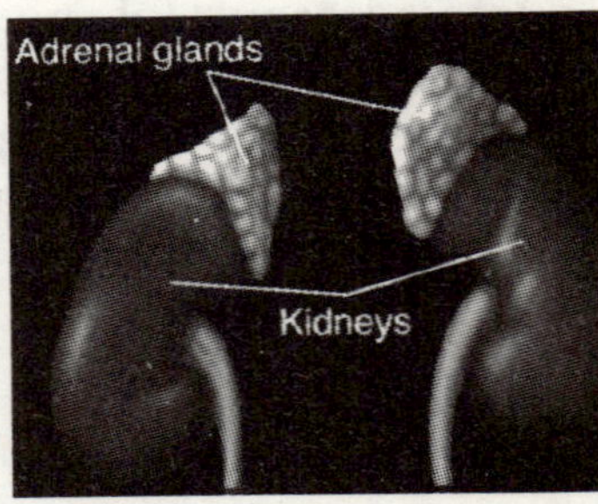

- **adrenergic**
See **sympathomimetic**

- **adsorbate**
the material being removed by the adsorption process.

- **adsorbent**
the material (activated carbon) that is responsible for removing the undesirable substance in the adsorption process.

- **adsorption**
transfer of gas or vapour molecules from the surrounding gas to a solid surface.

- **advanced wastewater treatment**
  any treatment of sewage that goes beyond the secondary or biological water treatment stage and includes the removal of nutrients such as phosphorus and nitrogen and a high percentage of suspended solids. (See **primary, secondary treatment.**)

- **adverse effect**
  change in morphology, physiology, growth, development or lifespan of an organism which results in impairment of functional capacity or impairment of capacity to compensate for additional stress or increase in susceptibility to the harmful effects of other environmental influences.

- **adverse event**
  occurrence which causes an adverse effect.

- **adverse health effect**
  a change in body function or cell structure that might lead to disease or health problems

- **aeration**
  the act of mixing a liquid with air (oxygen).

AERATION REACTORS

- **aerobe**
  organism which needs molecular oxygen for respiration and hence for growth and life.

- **aerobic**
  a biological process that occurs in the presence of oxygen.

- **aerobic biological oxidation**
  any waste treatment process utilising organisms in the presence of air or oxygen to reduce the pollution load or oxygen demand of organic substance in water.

- **aerobic digestion (sludge processing)**
  biochemical decomposition of organic matter by organisms living or active only in the presence of oxygen,

which results in the formation of mineral and simpler organic compounds.

- **aerobic organism**
  an organism that requires oxygen for its respiration.
- **aerobic treatment**
  process by which microbes decompose complex organic compounds in the presence of oxygen and use the liberated energy for reproduction and growth. Types of aerobic processes include extended aeration, trickling filtration, and rotating biological contactors.
- **aerodynamic (equivalent) diameter**
  diameter of a unit-density sphere having the same gravitational-settling velocity as the particle in question. Aerodynamic diameter

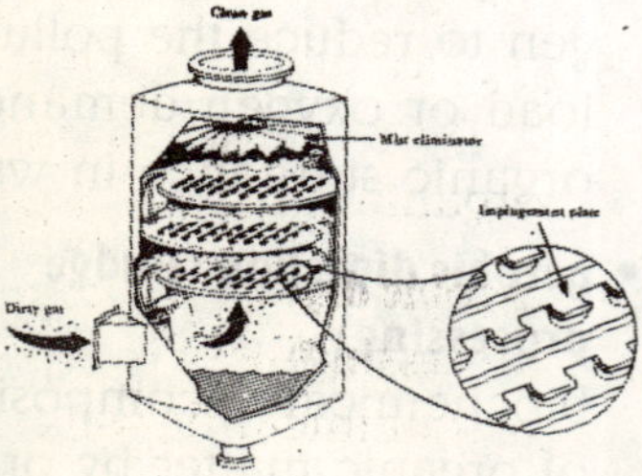

takes into account the shape, roughness; and aerodynamic drag of the particle. Used for movement of particles through a gas.

- **aerodynamic diameter**
  the diameter of a unit-density sphere having the same terminal settling velocity as the particle in question. It is used to predict where in the respiratory tract such particles will deposit.
- **aerodynamic particle sizer**
  a particle spectrometer that uses an acceleration system to differentiate particles by aerodynamic diameter and a laser velocimeter to detect particles.
- **aerosol**
  this is a very broad term applied to any suspension of solid or liquid particles in a gas. Particle diameters may range from about 0.001 micrometers to about 100 micrometers. Mass concentrations may range from 10-9 to 10 grams per cubic metre of gas.
- **aetiology**
  science dealing with the cause or origin of disease. In individuals, the cause or origin of disease. See **epidemiology**.

• **after-effect of a poison**
ability of a poison to produce a change in an organism after cessation of contact.

• **agar**
a medium used to grow micro-organisms.

• **age sensitivity**
quantitative and qualitative age dependence of an effect.

• **agglomerate**
a group of particles held together by van der Waals forces or surface tension.

• **agglomeration**
the process by which precipitation particles grow larger by collision or contact with cloud particles or other precipitation particles.

• **agonist**
substance which binds to cell receptors normally responding to naturally occurring substances and which

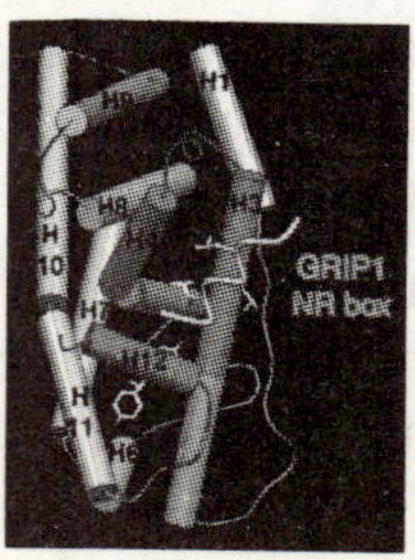

produces a response of its own. See **antagonist**.

• **agricultural waste**
poultry and livestock manure, and residual materials in liquid or solid form generated from the production and marketing of poultry, livestock, fur-bearing animals, and their products. Also includes grain, vegetable, and fruit harvest residue.

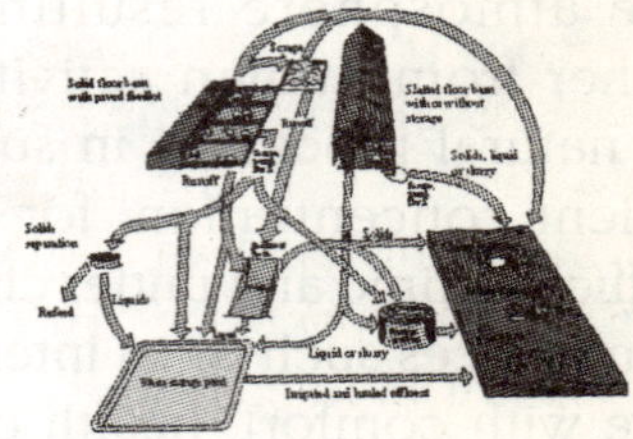

• **AIDS**
Acquired Immunodeficiency Syndrome. A condition in which the normal immune system becomes depressed, rendering the affected individual unable to fight a number of serious and fatal infections. Human T-cell leukaemia/ lymphoma virus (HTLV-III) is associated with AIDS. The virus has been renamed human immunodeficiency virus (HIV). The presence of

HIV antigen in the blood will confirm that an individual is infected with the virus and has AIDS. The presence of HIV antibody in the blood confirms that an individual has been infected with the AIDS virus. See **antigen and antibody**.

- **AIDS-related complex**
See **ARC**.

- **air pollution**
presence of substances in the atmosphere resulting either from human activity or natural processes, in sufficient concentration, for a sufficient time and under circumstances such as to interfere with comfort, health or welfare of persons or to harm the environment. See **pollution.**

- **air pollution control system**
1. network of organisations which monitor air pollution. 2. group of measures or processes used to minimise or prevent air pollution. See **air pollution**, **pollution.**

- **air quality standards**
the level of selected pollutants set by law that may not be exceeded in outside air. Used to determine the amount of pollutants that may be emitted by industry.

- **air stripping**
a treatment process used to remove dissolved gases and volatile substances from water. Large volumes of air are bubbled through the water being treated to remove (strip out) the dissolved gases and volatile substances. Also see **packed tower aeration**.

- **alar**
trade name for daminozide, a pesticide that makes apples redder, firmer, and less likely to drop off trees before growers are ready to pick them. Alar also is used to a lesser extent on peanuts, tart cherries, concord grapes, and other fruits.

- **albumin**
the most abundant protein component of blood. Produced primarily in the

liver, albumin helps to keep the fluid portion of blood within the vessels, and helps transport drugs and other substances by way of the bloodstream. Low levels of albumin in the blood or its presence in urine may signal oedema (the accumulation of fluid) as in pedal oedema (in the ankles) or pulmonary oedema (in the lungs), which may be symptoms of congestive heart failure, kidney or liver disease.

- **albuminuria**
  presence of albumin, derived from plasma, in the urine. See **microalbuminuria, proteinuria**.

- **algae**
  simple rootless plants that grow in sunlit waters in proportion to the amount of available nutrients. They can affect water quality adversely by lowering the dissolved oxygen in the water. They are food for fish and small aquatic animals.

- **algal blooms**
  sudden, massive growths of microscopic and macroscopic plant life, such as green or blue-green algae, which develop in lakes and reservoirs. They can affect water quality adversely and indicate potentially hazardous changes in local water chemistry.

- **algicide**
  substance intended to kill algae.

- **alkali**
  alkalis (or bases) are materials which have pH values from 8 to 14. Alkalis with pHs between 12 to 14 are considered to be corrosive and will cause severe dam-

age to skin and eyes. Any compound having highly basic properties; i.e., one that readily ionises in aqueous solution to yield OH^- anions, with a pH above 7, and turns

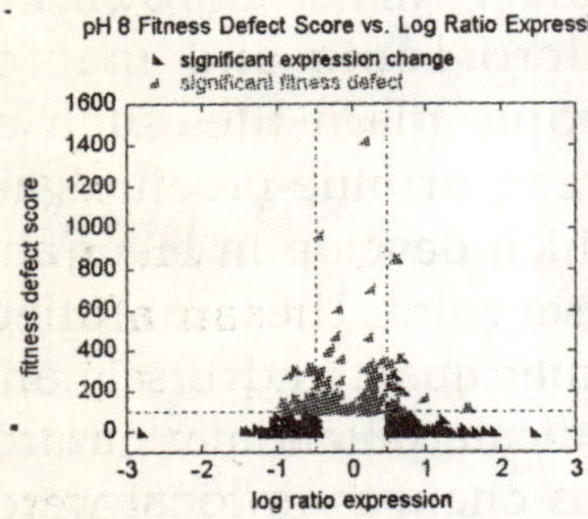

litmus paper blue. Alkalies are oxides and hydroxides of certain metals and belong to group IA of the periodic table (Li, Na, K, Rb, Cs, Fr). Ammonia and amines may also be alkaline. Alkalies are caustic and dissolve human tissue. Treat alkali burns by quickly washing the afflicted area with large amounts of water for at least 15 minutes. Common commercial alkalies are sodium carbonate (soda ash), NaOH, lime, lye, potash, caustic soda, KOH, water glass, regular mortar, portland cement, and bicarbonate of soda **(see Acid; pH)**.

- **alkaline**

having the properties of a base with a pH of more than 7. A common alkaline is baking soda.

- **alkaline phosphatase (Alk Phos)**

an enzyme found mainly in liver and bone. Blood serum alkaline phosphatase is usually measured to detect liver or bone disease.

- **alkalinity**

the capacity of water to neutralise acids. This capacity is caused by the water's content of carbonate, bicarbonate, hydroxide and occasionally borate, silicate, and phosphate. Alkalinity is expressed in milligrams per litre of equivalent calcium carbonate. Can be expressed as 'M' Alk to a methyl orange titration end point or 'P' Alk to a phenolphthalein end point. Alkalinity is not the same as pH because water does not have to be strongly basic (high pH) to have a high alkalinity. Alkalinity is a measure of how much acid can be added to a liquid without causing a great change in pH.

- **alkalosis**
  pathological condition in which the hydrogen ion substance concentration of body fluids is below normal and hence the pH of blood rises above the reference interval. See **acidosis**.

- **alkylating agent**
  this is a term that describes any substance that introduces an alkyl radical into a compound in place of a hydrogen atom.

EMS is an alkylating agent

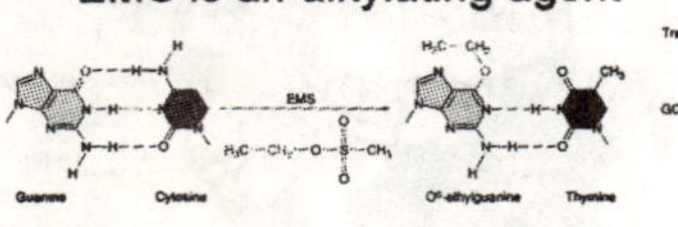

- **allele**
  one of several alternate forms of a gene which occur at the same relative position (locus) on homologous chromosomes and which become separated during meiosis and can be recombined following fusion of gametes. See **gametes, meiosis**.

- **allergen**
  antigenic substance capable of producing immediate hypersensitivity. See **allergy, antigen, hypersensitivity**.

- **allergy**
  symptoms or signs occurring in sensitised individuals following exposure to a previously encountered substance (allergen) which would otherwise not cause such symptoms or signs in non-sensitised individuals. The most common forms of allergy are rhinitis, urticaria, asthma, and contact dermatitis.

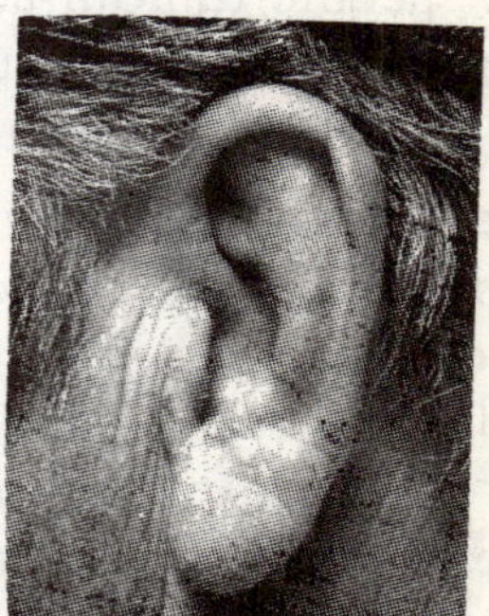

- **all-or-none effect**
  See **quantal effect, stochastic effect**.

- **alopecia**
baldness, absence or thinning of hair from areas of skin where it is usually present.

- **alpha-fetoprotein (AFP)**
a substance produced by the foetus that is found in foetal serum, amniotic fluid, and the mother's bloodstream. Elevated levels of AFP may indicate that the baby has a neural tube defect such as spina bifida (incomplete closure of the spinal column) which can lead to paralysis of the lower limbs, repeated urinary tract infections, mental retardation or hydrocephalus ('water on the brain'). It is also a useful nonspecific tumour-associated antigen (tumor marker).

- **alternative technology**
approach that aims to use resources efficiently or to substitute resources in order to do minimum damage to the environment. This approach permits a large degree of personal user control over the technology.

- **alveolar**
part of the respiratory system in which gas exchange occurs; alveoli are small sacs at the end of the bronchioles lined on the gas side by Types 1 and 2 epithelial cells.

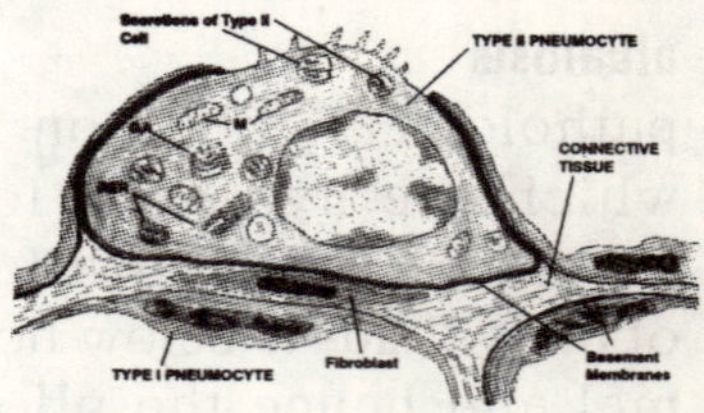

- **Alzheimer's disease**
a disorder marked by progressive mental deterioration (dementia), including memory loss, confusion, reduction of functional activity, and fluctuations in alertness, usually beginning

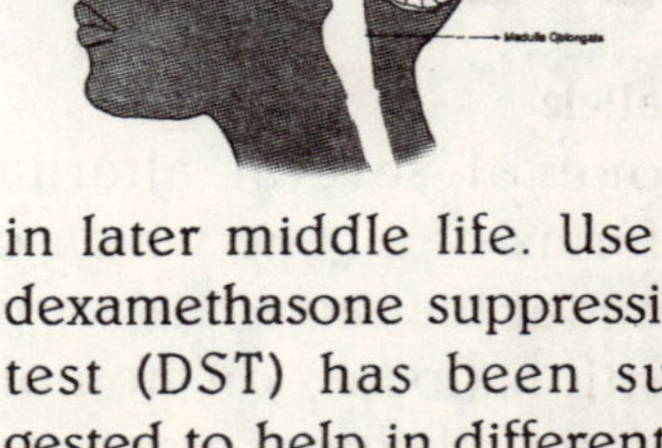

in later middle life. Use of dexamethasone suppression test (DST) has been suggested to help in differential diagnosis of dementia and depression (also a frequent disease of later life). This is important, as adequate treatment may produce improvement in depressive disease. See **dexamethasone suppression test**.

- **ambient**
surrounding (applied to environmental media such as air, water, sediment or soil).

- **ambient air**
surrounding air

- **ambient monitoring**
continuous or repeated measurement of agents in the environment to evaluate ambient exposure and health risk by comparison with appropriate reference values based on knowledge of the probable relationship between exposure and resultant adverse health effects. See **biological monitoring, environmental monitoring, monitoring**.

- **ambient standard**
See **environmental quality standard.**

- **ames test**
in vitro test for mutagenicity using mutant strains of the bacterium *Salmonella typhimurium* which cannot grow in a given histidine-deficient medium. Mutagens can cause reverse mutations which enable the bacterium to grow on the medium. The test can be carried out in the presence of a given microsomal fraction (S-9) from rat liver to allow metabolic transformation of mutagen precursors to active' derivatives.

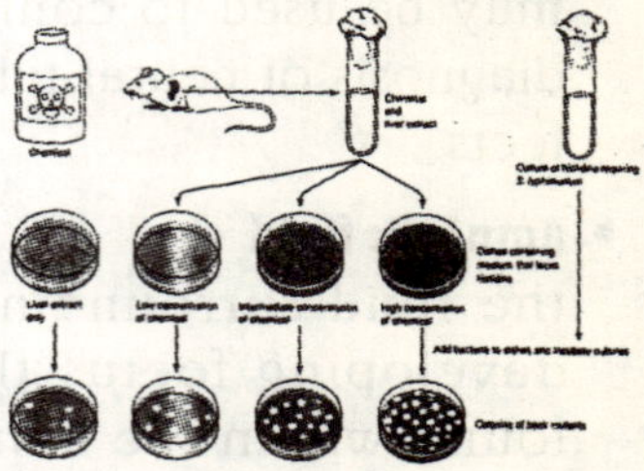

- **aminoglycosides**
a class of antibiotics specific for gram negative bacteria.

- **amniocentesis**
a prenatal diagnostic procedure in which a needle is inserted through the mother's abdomen into the uterus and amniotic sac to remove a sample of amniotic fluid. The fluid, containing foetal cells, proteins, and other substances, may be analysed to detect genetic disorders, including

Down syndrome (mongolism). An elevated level of AFP can indicate neural tube defects. See **neural tube defect.**

- **amniography**
  a prenatal diagnostic procedure in which X-ray of the mother's uterus, after injection of a radiopaque substance, allows visualisation of the foetus, the placenta, and the uterine lining. This may be used to confirm a diagnosis of neural tube defects.

- **amniotic fluid**
  the fluid surrounding the developing foetus that is found within the amniotic sac contained in the mother's womb.

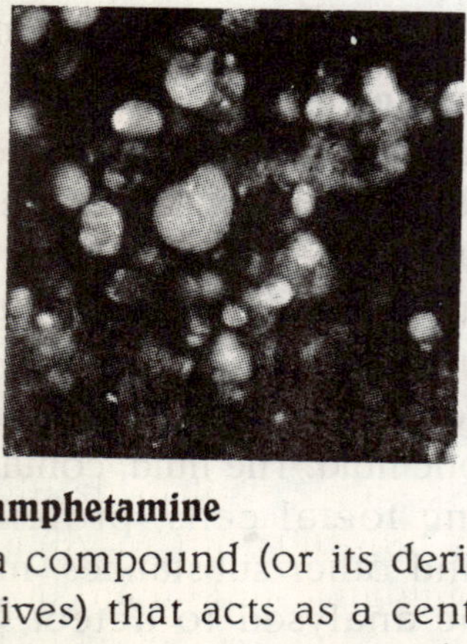

- **amphetamine**
  a compound (or its derivatives) that acts as a central nervous system stimulant. Used as a stimulant in depressed states, to reduce appetite, and in attention-deficit disorders, it can also alleviate nasal congestion. Amphetamine use can lead to dependence. It has a high potential for abuse. Blood and urine tests can detect amphetamine use and toxic levels.

- **amplicon**
  the product of PCR or LCR, a piece of DNA that has been synthesised using amplification techniques.

- **amplification (of genes)**
  See **gene amplification.**

- **amylase**
  1. a digestive enzyme produced largely by the pancreas and salivary glands that converts starches to sugars. Abnormally high levels of amylase in the blood or urine may be found in patients with inflammation of the pancreas or salivary glands (mumps).
  2. an organic chemical used as a cleaning solvent that causes adverse health effects in domestic water supplies.

• **trihalomethane**
one of a family of organic compounds named as derivatives of methane. THMs are generally the by-product from chlorination of drinking water that contains organic material. The resulting compounds (THMs) are suspected of causing cancer.

• **anabolism**
biochemical processes by which smaller molecules are joined to make larger molecules. See **catabolism.**

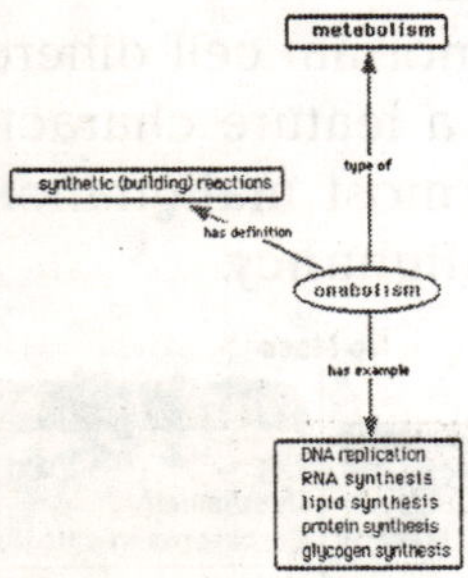

• **anaemia**
condition in which there is a reduction in the number of red blood cells or amount of haemoglobin per unit volume of blood below the reference interval for a similar individual of the species under consideration, often causing pallor and fatigue.

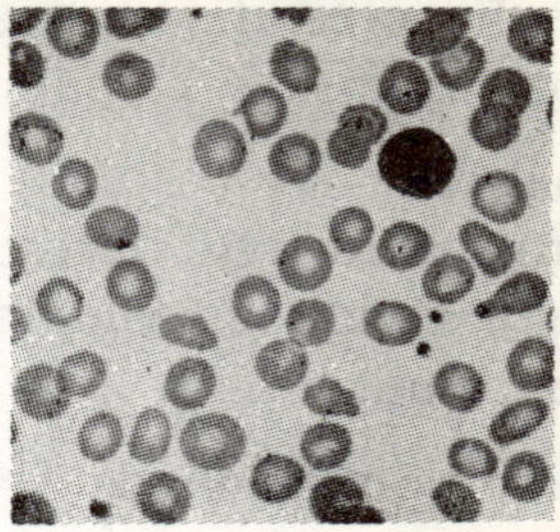

• **anaerobe**
organism which does not need molecular oxygen for life. Obligate (strict) anaerobes grow only in the absence of oxygen. Facultative anaerobes can grow either in the presence or in the absence of molecular oxygen. See **aerobe**.

• **anaerobic**
not requiring molecular oxygen.

• **anaerobic decomposition**
reduction of the net energy level and change in chemical composition of organic matter caused by micro-organisms in an oxygen free environment.

• **anaesthetic**
substance which produces loss of feeling or sensation. General anaesthetic produces loss of consciousness while local or regional an-

aesthetic renders a specific area insensible to pain.

- **analgesic**
substance which relieves pain, without causing loss of consciousness.

- **analogue metabolism**
process by which a normally non-biodegradable compound is biodegraded in the presence of a structurally similar compound which can induce the necessary enzymes.

- **analyte**
a substance measured in the laboratory. A chemical for which a sample (such as water, air, or blood) is tested in a laboratory. For example, if the analyte is mercury, the laboratory test will determine the amount of mercury in the sample.

Dual Analyte Assay Protocol

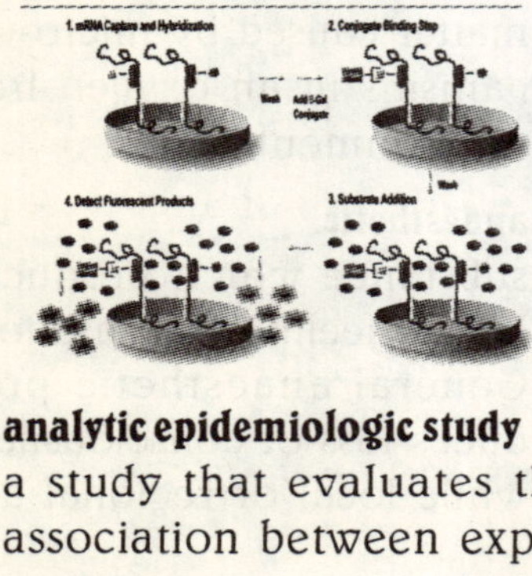

- **analytic epidemiologic study**
a study that evaluates the association between exposure to hazardous substances and disease by testing scientific hypothesis. Three types of analytic study are: cross-sectional (prevalence), cohort (prospective), and case control (retrospective).

- **anaphylaxis**
severe allergic reaction occurring in a person or animal exposed to an antigen or hapten to which they have previously been sensitised. See **antigen, hapten**.

- **anaplasia**
loss of normal cell differentiation, a feature characteristic of most malignancies. See **malignancy**.

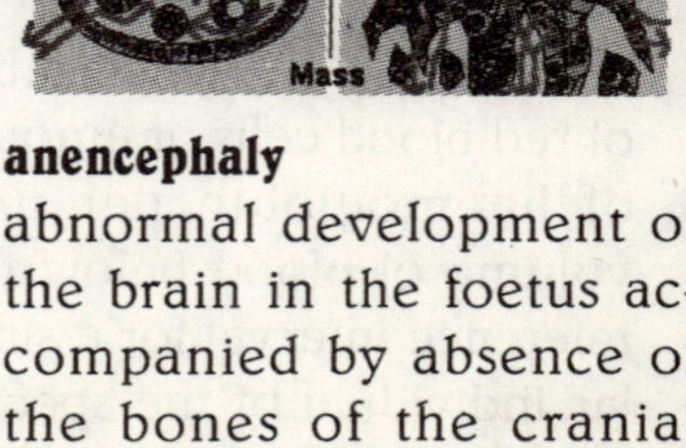

- **anencephaly**
abnormal development of the brain in the foetus accompanied by absence of the bones of the cranial vault. Children born with this severe disorder die

shortly after birth. Measurement of maternal blood and amniotic fluid levels of alpha-foetoprotein (AFP) can help diagnose this condition and other neural tube defects. See **alpha-foetoprotein**.

- **aneuploid**
cell or organism with missing or extra chromosomes or parts of chromosomes.

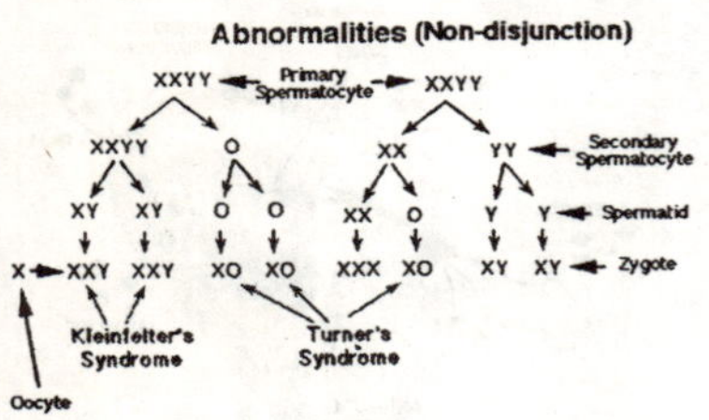

- **aneuploidy**
this is the condition of having an abnormal number of chromosomes for the species to which the organism under consideration belongs.

- **anisoaxial sampling**
sampling conditions in which the air flowing into an inlet has a different direction from the ambient air flow.

- **anova**
analysis of variance. A parametric method used for hypothesis testing (i.e. to determine if statistically significant differences in a response occur among two or more groups).

- **anoxia**
strictly total absence of oxygen but sometimes used to mean decreased oxygen supply in tissues.

- **antagonism**
combined effect of two or more factors which is smaller than the solitary effect of any one of those factors. In bioassays, the term may be used when a specified response is produced by exposure to either of two factors but not by exposure to both together.

- **antagonist**
substance that reverses or reduces the effect induced by an agonist. Substance that attaches to and blocks cell receptors that normally bind naturally occurring substances. See **agonist**.

- **antagonistic effect**
a biologic response to exposure to multiple substances that is less than would be

expected if the known effects of the individual substances were added together

- **anthelmint(h)ic**
substance intended to kill parasitic intestinal worms, such as helminths. See **antihelminth**.

- **anthracosis (coal miners' pneumoconiosis)**
form of pneumoconiosis caused by accumulation of carbon deposits in the lungs due to inhalation of smoke or coal dust.

- **anthropogenic**
caused by or influenced by human activities.

- **anti-adrenergic**
See **sympatholytic**.

- **antiarrhythmic**
any one of a group of drugs which prevent or control irregularities in the heart beat.

- **antiasthmatic**
any one of a group of drugs which prevent or control spasms and narrowing of the lung's bronchi and bronchioles.

- **antibiotic**
substance produced by, and obtained from, certain living cells (especially bacteria, yeasts and moulds), or an equivalent synthetic substance, which is biostatic or biocidal at low concentrations to some other form of life, especially pathogenic or noxious organisms.

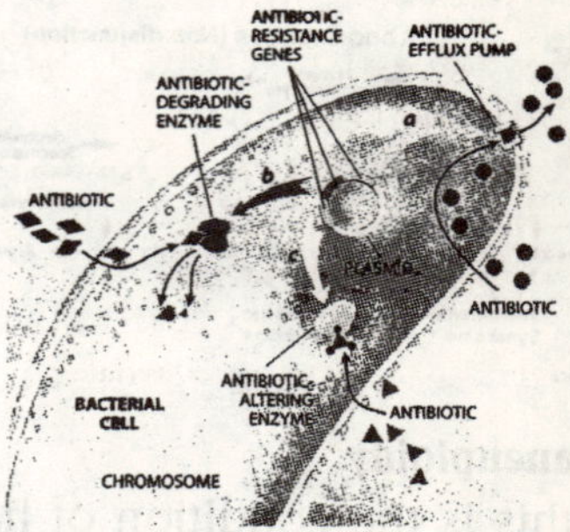

- **antibody**
protein molecule produced by the immune system (an immunoglobulin molecule) which can bind specifically to the molecule (antigen or hapten) which induced its synthesis. See **antigen, hapten, immunoglobulin**.

- **antibody assays**
tests to determine levels of specific antibodies in the blood.

• **anticholinergic**
1. preventing transmission of parasympathetic nerve impulses.
2. substance which prevents transmission of parasympathetic nerve impulses.

• **anticholinesterase**
See **cholinesterase inhibitor**.

• **anti-coagulant**
substance which prevents clotting. Examples: EDTA, heparin, sodium citrate, sodium oxalate.

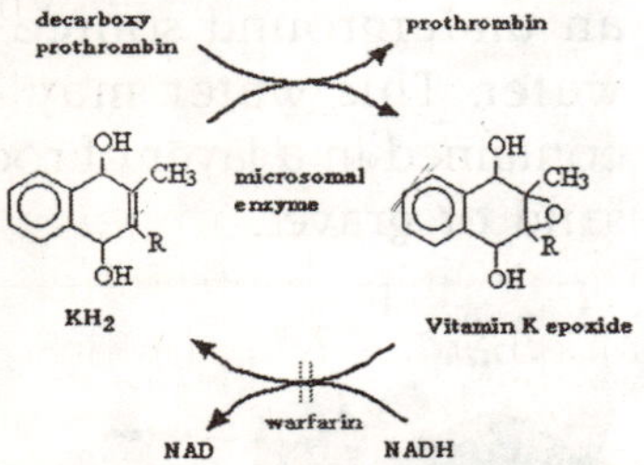

• **anti-convulsant**
any of a group of drugs that prevent seizures or control their incidence or severity.

• **antidote**
substance capable of specifically counteracting or reducing the effect of a potentially toxic substance in an organism by a relatively specific chemical or pharmacological action.

• **antigen**
substance or a structural part of a substance which causes the immune system to produce specific antibody or specific cells and which combines with specific binding sites (epitopes) on the antibody or cells. See **antibody**, **epitope**.

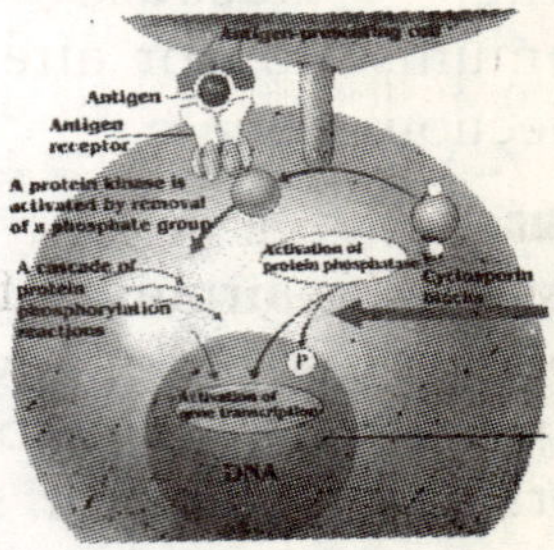

• **anti-helminth**
See **anthelmint(h)ic**.

• **antimetabolite**
substance, structurally similar to a metabolite, which competes with it or replaces it, and so prevents or reduces its normal utilisation.

• **antimycotic**
substance used to kill a fungus or to inhibit its growth. See **fungicide**.

• **antipyretic**
substance which relieves or reduces fever.

- **anti-resistant**
substance used as an additive to a pesticide formulation in order to reduce the resistance of insects to the pesticide.

- **antiserum**
serum containing antibodies to a particular antigen either because of immunisation or after an infectious disease.

- **aphasia**
loss or impairment of the power of speech or writing, or of the ability to understand written or spoken language or signs, due to a brain injury or disease.

- **aphicide**
substance intended to kill aphids. See **insecticide.**

- **aphid**
common name for a harmful plant parasite in the family Aphididae. Some species are vectors of plant virus diseases.

- **aplasia**
lack of development of an organ or tissue, or of the cellular products from an organ or tissue.

- **apoptosis**
physiological process of programmed tissue death (and disintegration) associated with normal development in animals. See **necrosis.**

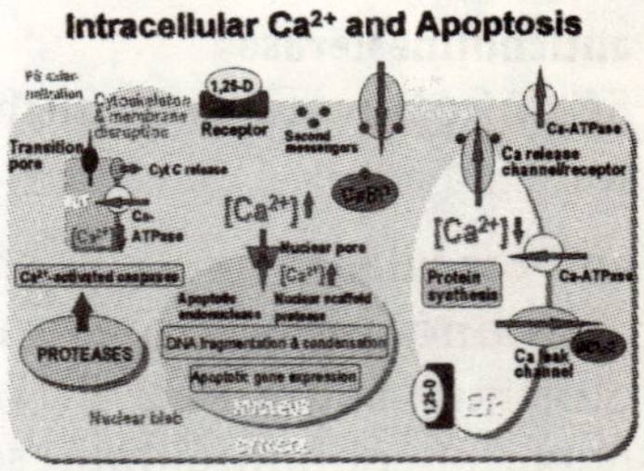

- **aquifer**
an underground source of water. This water may be contained in a layer of rock, sand or gravel.

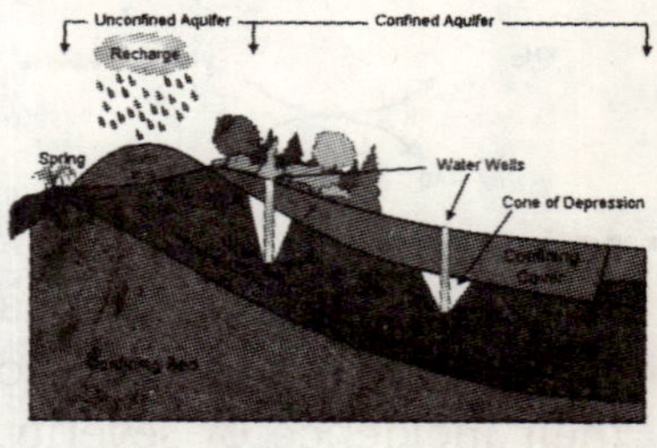

- **arboricide**
substance intended to kill trees and shrubs.

- **ARC**
AIDS-related complex. A condition in which antibody tests for AIDS virus are positive and patients may exhibit enlarged

lymph nodes, fatigue, fever, night sweats, weight loss, and unexplained diarrhoea, but do not have any of the more serious complications of AIDS.

- **area source**
widespread origin of emissions. See **point source**.

- **argyria**
pathological condition characterised by grey-bluish or black pigmentation of tissues (such as skin, retina, mucous membranes, internal organs) caused by the accumulation of metallic silver, due to reduction of a silver compound which has entered the organism during (prolonged) administration or exposure. See **argyrosis**.

- **argyrosis**
See **argyria**.

- **arrhythmia**
any variation from the normal rhythm of the heartbeat.

- **artefact**
finding or product of experimental or observational techniques that is not properly associated with the system being studied.

- **arteriogram**
X-ray visualisation of the arterial lining after injection of radiopaque dye into a specific artery or into the bloodstream. It aids in the diagnosis of vascular occlusions and of athrosclerotic plaques, as well as of other cardiovascular abnormalities.

- **arteriosclerosis**
hardening and thickening of the walls of the arteries.

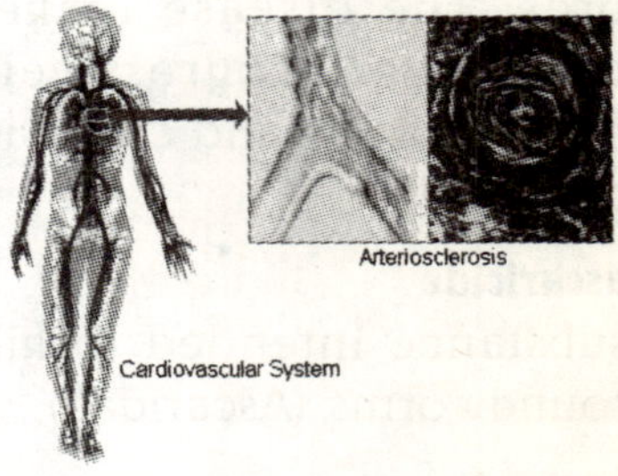

- **arthralgia**
pain in a joint.

- **arthralgia saturnia**
pain in a joint resulting from lead poisoning.

- **arthritis**
inflammation of a joint, usually accompanied by pain and often by changes in structure.

• **asbestos**
a mineral fibre that can pollute air or water and cause cancer or asbestosis when inhaled.

• **asbestosis**
form of pneumoconiosis caused by inhalation of asbestos fibres. See **pneumoconiosis.**

• **asbestosis**
a disease associated with chronic exposure to and inhalation of asbestos fibres. The disease makes breathing progressively more difficult and can lead to death.

• **ascaricide**
substance intended to kill roundworms (Ascaridae).

• **asphyxia**
1. condition resulting from insufficient intake of oxygen. Symptoms include breathing difficulty, impairment of senses, and, in extreme, convulsions, unconsciousness and death.
2. lack of oxygen and interference with the oxygenation of the blood. Can lead to unconsciousness.

• **asphyxiant**
a vapour or gas that can cause unconsciousness or death by suffocation (lack of oxygen). Most simple asphyxiants are harmful to the body only when they become so concentrated that they reduce (displace) the available oxygen in the air (normally about 21%) to dangerous levels (18% or lower). Examples of simple asphyxiants are CO_2, N_2, H_2, and He. Chemical asphyxiants like carbon monoxide (CO) reduce the blood's ability to carry oxygen, or like cyanide interfere with the body's utilisation of oxygen.

• **asphyxiation**
a condition that causes asphyxia or suffocation. Asphyxiation is one of the principal potential hazards of working in confined spaces.

• **aspiration hazard**
the danger of drawing material into the lungs, leading to an inflammatory response that can be fatal.

- **assay**
analysis of a substance, or of components in a mixture, or of the biological, chemical, or pharmacological potency of a drug. Forms of assay include: endpoint assay — a single measurement is made at a fixed time, kinetic assay — increasing amounts of a product are formed with time, and are monitored at multiple points, microbiological assay — measurement of the concentration of antimicrobials in biological material, immunological assay — analysis or measurement is based on antigen-antibody reactions.

- **assay (indicator-labelled immunoassays)**
indicators are attached to the antigen or to the antibody to demonstrate that the antigen-antibody reaction has occurred. These assays have enabled measurement of many substances not previously measurable. Among these assays are: Enzyme immunoassay (EIA), Enzyme-linked immunosorbent assay (ELISA), Enzyme-multiplied immunoassay technique (EMIT), Fluorescence immunoassay (FIA), Fluorescence polarisation immunoassay (FPIA) and Radio immunoassay (RIA).

- **assessment of exposure**
See **biological assessment of exposure.**

- **assimilation**
the ability of a body of water to purify itself of pollutants.

- **assimilative capacity**
the ability of a natural body of water to receive wastewaters or toxic materials without harmful effects and without damage to aquatic life.

- **asthenia**
weakness, lack or loss of strength.

- **asthma**
chronic respiratory disease characterised by, bronchoconstriction, excessive mucous secretion and oedema of the pulmonary alveoli, resulting in difficulty in breathing out, wheezing, and cough.

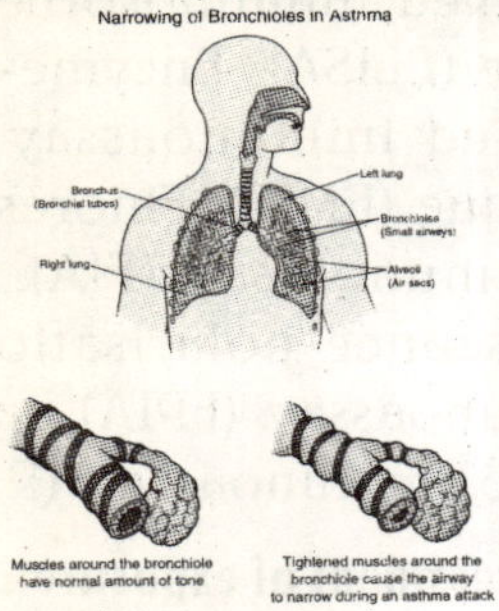

- **astringent**
causing contraction, usually locally after topical application. Substance causing cells to shrink, thus causing tissue contraction or stoppage of secretions and discharges. Such substances may be applied to skin to harden and protect it.

- **ataxia**
unsteady or irregular manner of walking or movement caused by loss or failure of muscular coordination.

- **atherosclerosis**
pathological condition in which there is thickening, hardening, and loss of elasticity of the walls of blood vessels, characterised by a variable combination of changes of the innermost layer consisting of local accumulation of lipids, complex carbohydrates, blood and blood components, fibrous tissue and calcium deposits. In addition, the outer layer becomes thickened and there is fatty degeneration of the middle layer.

- **atmosphere**
1. a standard unit of pressure representing the pressure exerted by a 29.92 inch column of mercury at sea level at 45 degrees latitude and equal to 1000 grams per square centimetre.
2. the whole mass of air surrounding the earth, composed largely of oxygen and nitrogen.

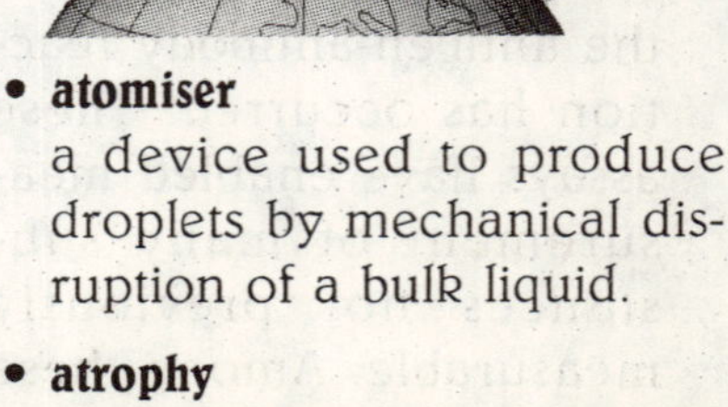

- **atomiser**
a device used to produce droplets by mechanical disruption of a bulk liquid.

- **atrophy**
wasting away of the body or of an organ or tissue.

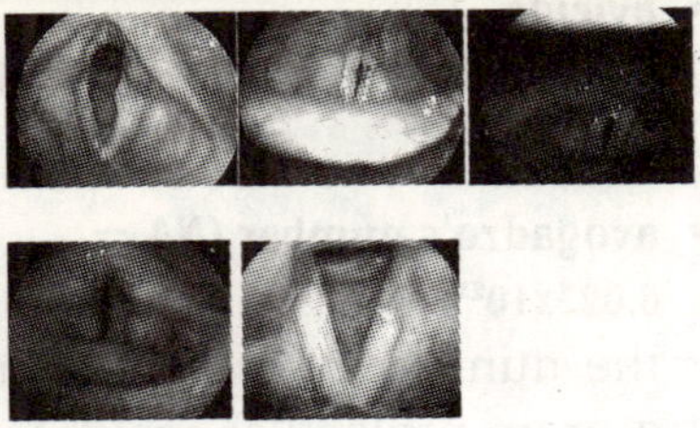

- **attenuation**
  1. a decrease in light due to absorption by a chromagen in radiative energy attenuation technologies.
  2. regulation of gene expression in bacteria by premature termination of transcription of a biosynthetic operon.
- **attractant**
  substance used to attract animals with the aim of killing or sterilising them. See **pheromone.**
- **attributable risk**
  difference between the risk of exhibiting a certain adverse effect in the presence of a substance and the same risk in the absence of the substance. See **risk**.
- **autoimmune disease**
  pathological condition resulting when an organism produces antibodies or specific cells which bind to constituents of its own tissues (autoantigens) and cause tissue injury. Examples of such diseases may include rheumatoid arthritis, myasthenia gravis, and scleroderma.

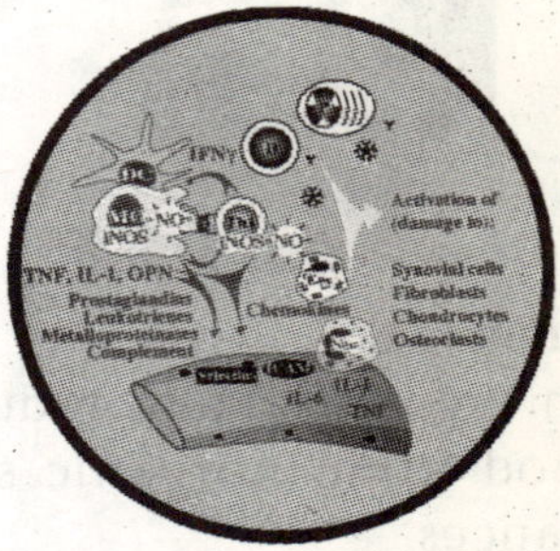

See **allergy, antibody, antigen, hypersensitivity, immune response**.

- **autoimmune(autoimmunity)**
  a condition characterised by a specific humoral or cell-mediated immune response against constituents of the body's own tissues.
- **autophagosome**
  membrane-bound body (secondary lysosome) in which parts of the cell are digested.
- **autopsy**
  post-mortem examination of the organs and body tissue to determine cause of death or pathological con-

dition. See **biopsy**, **necropsy**.

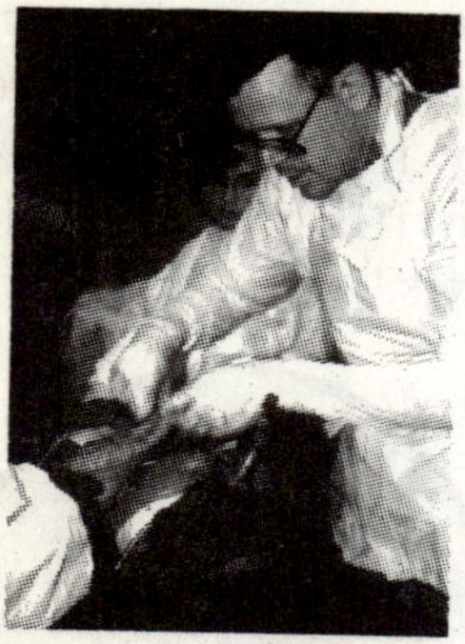

- **autotrophic**
  an organism that produces food from inorganic substances.

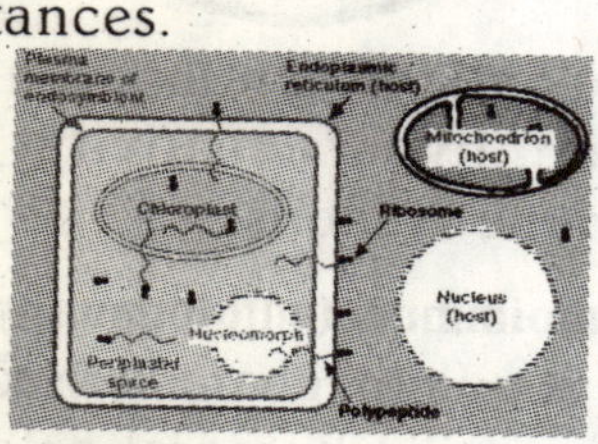

- **auxotroph**
  organism unable to synthesise an organic molecule which is required for its growth. When the compound is given to the organism with the other nutrients it requires, growth of the organism may occur.

- **auxotrophy**
  inability of a micro-organism to synthesise a particular organic compound required for its growth.

- **avicide**
  substance intended to kill birds.

- **avogadro's number (NA = 6.023×10^{23})**
  the number of molecules in a gram-molecular weight of any substance.

- **axenic animal**
  see **germ free animal**.

- **B lymphocyte**
  type of lymphocyte which synthesises and secretes antibodies in response to the presence of a foreign substance or one identified by it as foreign. The protective effect can be mediated to a certain extent by the antibody alone (contrast T lymphocyte). See **immune response, lymphocyte, T lymphocyte**.

- **background level**
  an average or expected amount of a substance or radioactive material in a specific environment, or typical amounts of substances that occur naturally in an environment.

- **back-mutation**
  process which reverses the effect of a mutation which

had inactivated a gene. Thus it restores the wild phenotype. See **phenotype.**

- **bacteria**
singular: bacterium. Microscopic living organisms usually consisting of a single cell. Bacteria can aid in pollution control by consuming or breaking down organic matter in sewage, or by similarly acting on oil spills or other water pollutants. Some bacteria in soil, water or air may also cause human, animal and plant health problems.

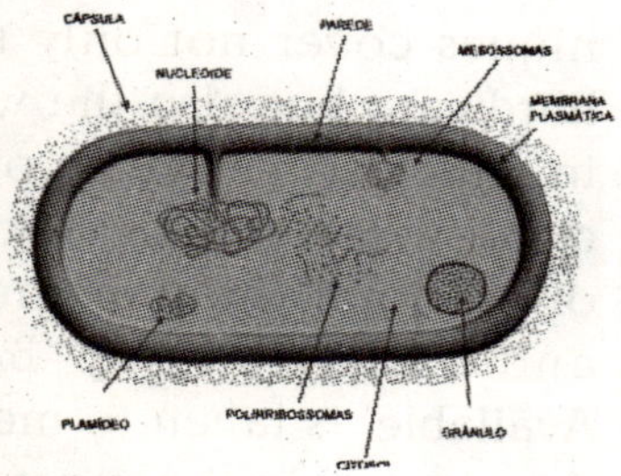

- **bactericide**
substance intended to kill bacteria.

- **bagassosis**
lung disease caused by the inhalation of dust from sugar-cane residues.

- **barbiturates**
a group of drugs derived from barbituric acid that is used to sedate, to control convulsions, or to induce sleep. Blood and urine tests can determine toxic levels of these compounds. Barbiturates may be habit forming.

- **base pairing**
linking of the complementary pair of polynucleotide chains of nucleic acids by means of hydrogen bonds between complementary purine and pyrimidine bases, adenine with thymine or uracil, cytosine with guanine.

- **batch testing**
a laboratory testing procedure in which one test is done simultaneously on multiple specimens in a 'batch analyser', such as a centrifugal analyser.

- **B-cell**
See **B lymphocyte.**

- **bedrock**
the solid rock underneath surface soils.

- **benefit**
advantage to or improvement in condition of an individual or a population.

• **benign**
1. of a disease, producing no persisting harmful effects.
2. tumour which does not invade other tissues (metastasise), having lost growth control but not positional control. See **malignant**.

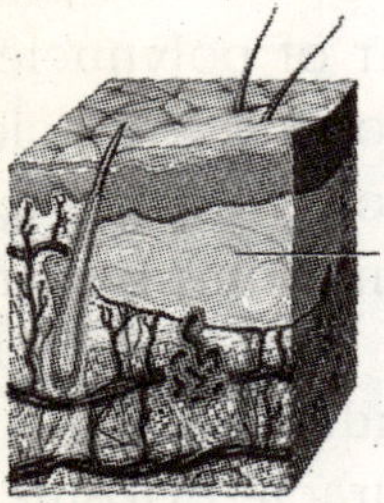

• **benthic organism**
any of a diverse group of aquatic plants and animals that lives on the bottom of marine and fresh bodies of water. The presence or absence of certain benthic organisms can be used as an indicator of water quality.

• **benzodiazepine**
a chemical that is often used as the active ingredient in mind-affecting drugs such as tranquillisers. Long term use can result in dependence. Benzodiazepine overdose can be determined through blood or urine tests.

• **berylliosis**
see **beryllium disease**

• **beryllium disease**
serious and usually permanent lung damage resulting from chronic inhalation of beryllium.

• **Best Available Techniques Not Entailing Excessive Costs (BATNEEC)**
this is essentially an updating and refinement of Best Practicable Means (BPM) whereby a greater degree of control over emissions to land, air and water may be exercised. The techniques cover not only the hardware but also the way in which a particular process is operated. In fact, as operational management and staff training 'Best Available' is taken to mean demonstrably the most effective techniques for an operation at the appropriate scale and commercial availability even if they are not in general use. This is a most important development from the previous BPM philosophy. 'Not Entailing excessive costs' means that the benefits

gained by using the best available techniques should be at a reasonable relationship to the cost of obtaining them, otherwise it may not be justifiable to obtain very small reductions in emissions at very high costs unless emissions are very toxic, in which case very high specifications may be imposed, e.g. as in hazardous waste incineration. Judgement about what constitutes BATNEEC can differ between new and existing plants. For example, the cost of adopting given techniques at an existing plant can depend upon its technical characteristics and could be much higher than those at a new plant which is designed to incorporate these techniques from the beginning. The benefits from adopting these techniques could also be less at an existing plant, depending on the remaining life and likely utilisation, in other words a degree of flexibility is necessary.

- **Best Practicable Environmental Option (BPEO) Index**

a numerical estimate of the overall environmental input of a range of feasible pollution control options across the media of air, land, and water. The options can then be compared.

- **Best Practicable Means (BPM)**

an essentially pragmatic philosophy for the control of emissions from scheduled processes.

- **bias**

deviation of results or inferences from the truth, or processes leading to such deviation. Any trend in the collection, analysis, interpretation, publication, or review of data which can lead to conclusions which are systematically different from the truth. Among the ways in which deviation from the truth can occur are the following:

1. systematic (one-sided) variation of measurements from the true values. See **systematic error**.

2. variation of statistical summary measures (means, rates, measures of associa-

tion, etc.) from their true values as a result of systematic variation of measurements, other flaws in data collection, or flaws in study design or analysis.

3. deviation of inferences from the truth as a result of flaws in study design, data collection, or the analysis or interpretation of results.

4. a tendency of procedures (in study design, data collection, analysis, interpretation, review or publication) to yield results or conclusions which depart from the truth.

5. prejudice leading to the conscious or unconscious selection of study procedures which depart from the truth in a particular direction, or to one-sidedness in the interpretation of results.

- **biased sample**

any sample which is not a random sample. See **random sample**, **sample**, **stratified sample**, **systematic sample**.

- **bicarbonate**

essential for regulating vital functions and one of the important buffers necessary to maintain normal acid-base balance in the body. Body metabolism results in mainly acid production, and neutralising some of such acids is its constant activity. Thus it plays a key role in metabolic acidosis or alkalosis. Bicarbonate may be lost through watery faeces, or can decrease when lungs cannot expel carbon dioxide.

- **bichromatics**

spectrophotometry which uses a secondary wavelength absorbance reading subtracted from a primary wavelength absorbance reading to obtain a delta absorbance reading (Ad). The delta absorbance is used in conjunction with the calibration data to calculate concentration or activity.

- **bi-directional**

refers to communication between an instrument and a computer that allows the computer to send instructions to the instrument and for the instrument to send results back to the computer.

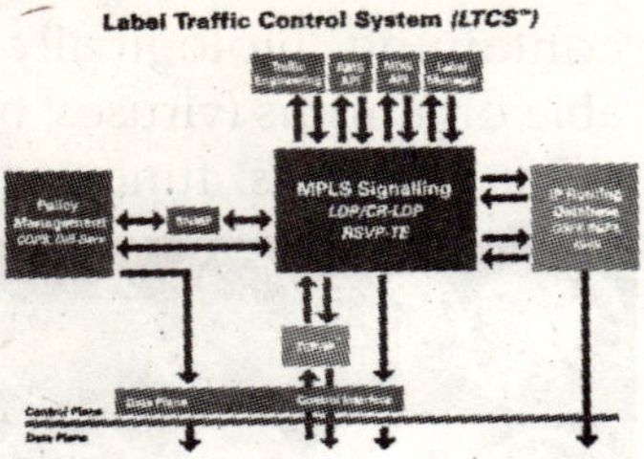

- **bilirubin**
the orange-yellow pigment of bile, the green fluid that aids in digestion and that is secreted by the liver. Bilirubin is formed mainly by haemoglobin breakdown at the end of red cell life and eventually most of it leaves the body in the faeces. Two types are in the blood. Water insoluble or unconjugated bilirubin refers to the pigment before it reaches the liver. In the liver it is converted to the water-soluble or conjugated bilirubin which is excreted into the bile. Blood tests for total bilirubin measure both unconjugated and conjugated bilirubin, and are performed to evaluate jaundice, anemia, various liver diseases (including hepatitis), and impaired bile excretion.

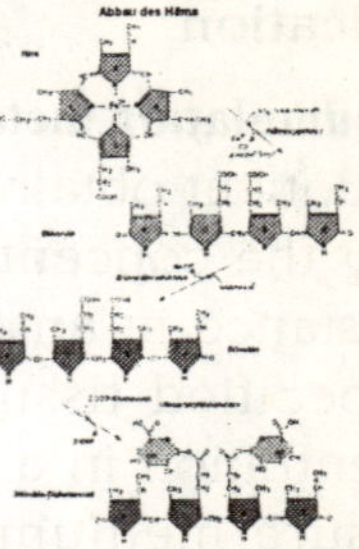

- **bimodal size distribution**
particle size distribution with two distinct maxima.

- **bioaccumulate**
1. process in which harmful substances move through the food chain, by one creature eating another, accumulating (at higher concentrations) in the top carnivores.

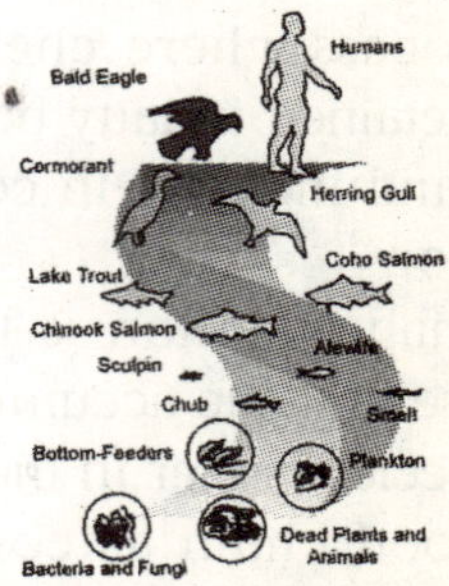

2. progressive increase in the amount of a substance in an organism or part of an organism which occurs because the rate of intake exceeds the organism's ability to remove

the substance from the body. See **bioconcentration**, **biom agnification**.

- **bioaccumulation factor (BAF)**
the quotient obtained by dividing the concentration of a substance in an organism (or specified tissue) by its concentration in a specified exposure medium, for example, air, food, sediment, soil, water, when several media are possible sources

- **bioaccumulation potential**
ability of living organisms to concentrate a substance obtained either directly from the environment or indirectly through its food.

- **bioaccumulation/ biomagnification**
a process where chemicals are retained in fatty body tissue and increase in concentration over time. Biomagnification is the increase of tissue accumulation in species higher in the natural food chain as contaminated food species are eaten.

- **bioactivation**
any metabolic conversion of a xenobiotic to a more toxic derivative. See **activation**, **bio-transformation**.

- **bioaerosol**
an aerosol of solid or liquid particles consisting of, or containing, biologically-viable organisms (viruses, bacteria, allergans, fungi, etc.),

with size ranging from sub-micrometer to greater than 100μm.

- **bioassay**
procedure for estimating the concentration or biological activity of a substance (vitamin, hormone, plant growth factor, antibiotic etc.) by measuring its effect on an organism compared to an appropriate standard preparation. See **assay**.

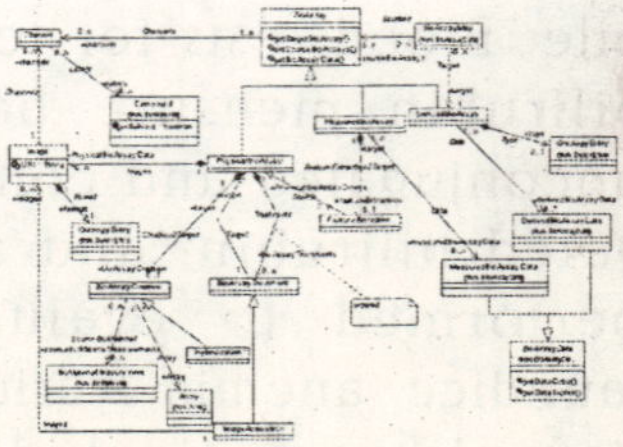

- **bioavailability**

1. extent to which a substance to which the body is exposed (by ingestion, inhalation, injection, or skin contact) reaches the systemic circulation, and the rate at which this occurs. See **biological availability, physiological availability.**

2. pharmacokinetic term relating systemic exposure from extravascular exposure (ev) to that following intravenous exposure (iv) by the equation:

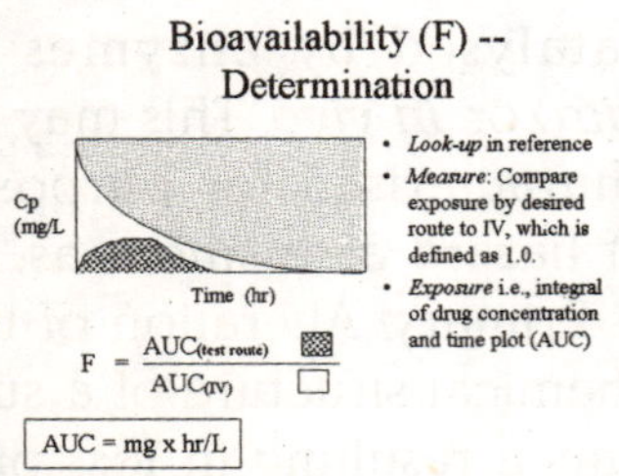

$$F = \mathrm{AUC}_{ev} * D_{iv} / \mathrm{AUC}_{iv} * D_{ev}$$

where F is the bioavailability, AUC_{ev} and AUC_{iv} are the areas under the plasma concentration time curve following extravascular and intravenous administration and D_{ev} and D_{iv} are the administered extravascular and intravenous doses. See **chemical oxygen demand.**

- **biochemical mechanism**

reaction or series of reactions, usually enzyme-catalysed, associated with a specific physiological event in a living organism.

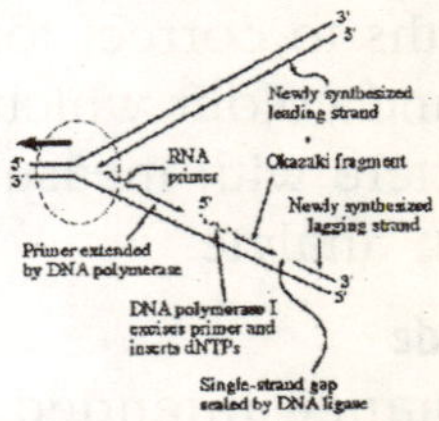

- **biochemical oxygen demand (BOD)**

a measure of the oxygen required to break down organic materials in water. Higher organic loads require larger amounts of oxygen and may reduce the amount of oxygen available for fish and aquatic life below acceptable levels.

- **biochemistry**

the chemistry of living organisms and their vital processes.

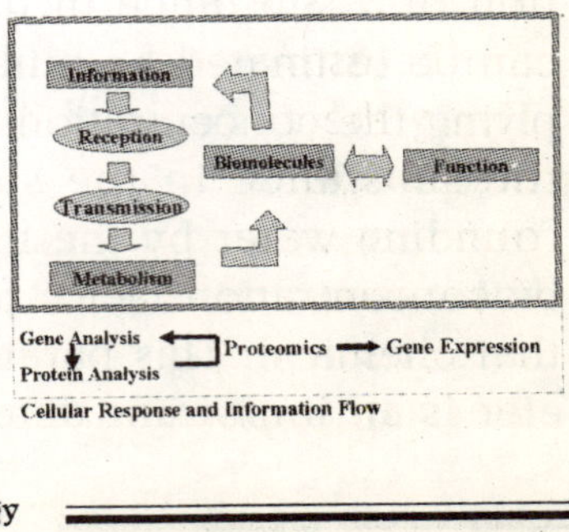

- **biochromatic chemical analyser**
laboratory instrumentation that provides spectrophotometric-photometric monitoring at two different wavelengths to correct for background colour which might interfere with measurement of an analyte.

- **biocide**
substance intended to kill living organisms.

- **bioconcentration**
process leading to a higher concentration of a substance in an organism than in environmental media to which it is exposed. See **bioaccumulation, biomagnification.**

- **bioconcentration factor (BCF)**
measure of the tendency for a substance in water to accumulate in fish tissue or in tissues of other organisms. The equilibrium concentration of a substance in fish can be estimated by multiplying the concentration of the substance in the surrounding water by the fish bioconcentration factor for that chemical. This parameter is an important determinant for human intake by the aquatic food ingestion route.

- **bioconversion**
see **bio-transformation**.

- **biodegradable**
the ability of a substance to be broken down physically and/or chemically by microorganisms. For example, many chemicals, food scraps, cotton, wool, and paper are bio-degradable, plastics and polyester generally are not.

- **biodegradation**
breakdown of a substance catalysed by enzymes *in vitro* or *in vivo*. This may be characterised for purposes of hazard assessment as:
1. Primary. Alteration of the chemical structure of a substance resulting in loss of a specific property of that substance.
2. Environmentally acceptable. Biodegradation to such an extent as to remove undesirable properties of the compound. This often corresponds to primary biodegradation but it depends on the circumstances under which the products are dis-

charged into the environment.

3. Ultimate. Complete breakdown of a compound to either fully oxidised or reduced simple molecules (such as carbon dioxide/methane, nitrate/ammonium, and water. It should be noted that the products of biodegradation can be more harmful than the substance degraded. See **bio-transformation**.

- **biodiversity**
  the number and variety of different organisms in the ecological complexes in which they naturally occur. Organisms are organised at many levels, ranging from complete ecosystems to the biochemical structures that are the molecular basis of heredity. Thus, the term encompasses different ecosystems, species, and genes that must be present for a healthy environment. A large number of species must characterise the food chain, representing multiple predator-prey relationships.

- **bio-elimination**
  removal, usually from the aqueous phase, of a test substance in the presence of living organisms by biological processes supplemented by physico-chemical reactions.

- **bio-equivalence**
  relationship between two preparations of the same drug in the same dosage form that have a similar bioavailability.

- **biological indicators of exposure study**
  a study that uses (1) biomedical testing or (2) the measurement of a substance [an analyte], its metabolite, or another marker of exposure in human body fluids or tissues to confirm human exposure to a hazardous substance [also see **exposure investigation**].

- **biological monitoring**
  measuring hazardous substances in biological materials (such as blood, hair, urine, or breath) to determine whether exposure has occurred. A blood test for lead is an example of biological monitoring.

- **biological uptake**
  the transfer of substances from the environment to plants, animals, and humans.

- **biological absorption**
  see **absorption, biological.**

- **biological acclimatisation**
  see **acclimatisation, biological.**

- **biological assessment of exposure**
  assessment of exposure to a substance by the analysis of specimens taken in the environment such as foodstuffs, plants, animals, biological material in air or water samples, or biological material from exposed subjects. When human samples are analysed, they are usually urine and blood. Other possible samples include expired air, faeces, saliva, bile, hair, and biopsy or autopsy material. When other organisms are being considered, the whole organism may be analysed as well as selected tissues such as fat in pigs or birds. In these samples, the content(s) of the substance(s) or metabolite(s) is determined and, on this basis, the exposure level (concentration in the air, absorbed amount of the substance) or the probability of health impairment due to exposure are derived. Biochemical changes in the components of an organism, such as changes in enzyme activity or in the excretion of metabolic intermediates, can also be used for this purpose if they show a relationship to the exposure. See **biological monitoring, monitoring.**

- **biological cycle**
  complete circulatory process through which a substance passes in the biosphere. It may involve transport through the various media (air, water, soil), followed by environmental transformation, and carriage through various ecosystems. See **biosphere, ecosystem.**

- **biological effect monitoring (BEM)**
  continuous or repeated measurement of early biological effects of exposure to a substance to evaluate ambient exposure and

health risk by comparison with appropriate reference values based on knowledge of the probable relationship between ambient exposure and biological effects. See **biological monitoring, environmental monitoring**.

- **biological half-life (T1/2)**

the term which is used for the time required for the amount of a particular substance in a biological system to be reduced to one half of its value by biological processes when the rate of removal is approximately exponential. Substances with a long biological half-life will tend to accumulate in the body and are, therefore, particularly to be avoided. Substances with a short biological half-life may accumulate if some becomes tightly bound, even if most is cleared from the body rapidly. There is also the possibility of cumulative effects of chemicals which have a short residence time in the body.

- **biological monitoring**

this is a procedure of periodic examination of biological specimens for the purposes of monitoring. It is usually applied to exposure monitoring but can also apply to effect monitoring. And analysis of the amounts of potentially toxic substances or their metabolites present in body tissues and fluids, as a means of assessing exposure to these substances and aiding timely action to prevent adverse effects. The term is also used to mean assessment of the biological status of populations and communities of organisms at risk, in order to protect them and to have early warning of possible hazards to human health.

- **biological oxidation**

the way that bacteria and micro-organisms feed on and decompose complex organic materials. Used in self-purification of water bodies and activated sludge wastewater treatment. See **biochemical oxygen demand**.

- **biological pesticides**
certain micro-organisms, including bacteria, fungi, viruses, and protozoa that are effective in controlling target pests. These agents usually do not have toxic effects on animals and people and do not leave toxic or persistent chemical residues in the environment.

- **biological preparation**
compound derived from living organisms and their products for use in medicine or as a pesticide etc. See **biological, bio-preparation.**

- **biological specimen**
1. organ, tissue (including blood), secretion or excretion product taken from an organism as a sample reflecting the state of the whole organism.
2. organism taken as a sample reflecting the state of a population or their environment.

3-D Electron Micrograph of Biological Specimen

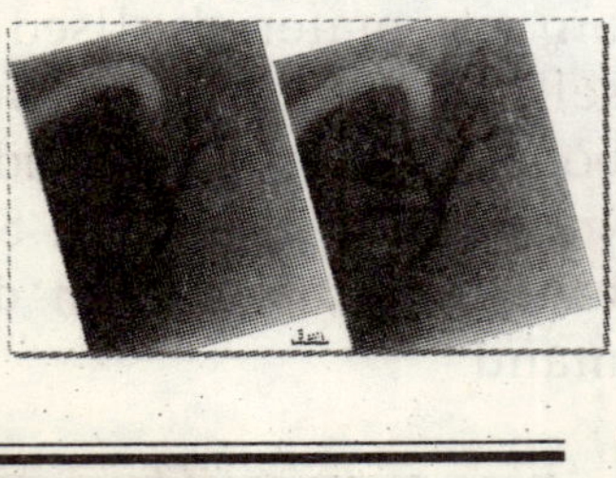

- **bio-magnification**
sequence of processes in an ecosystem by which higher concentrations are attained

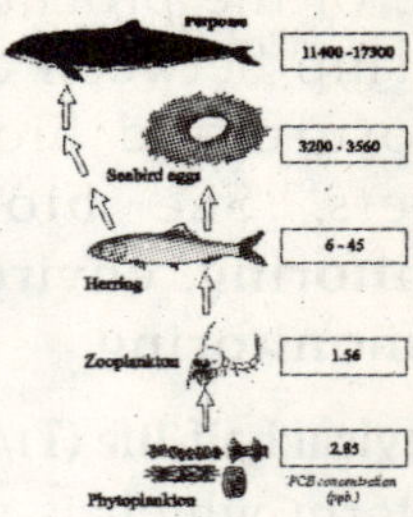

The concentration of PCB's tends to increase in the tissues of organisms at higher levels in the marine food chain, a phenomenon termed 'biomagnification'. [From: Percy, Wells and Evans [illegible], 1996: see Further Reading]

in organisms at higher trophic levels (at higher levels in the food web). At its simplest, it's a process leading to a higher concentration of a substance in an organism than in its food. See **ecological magnification, bioaccumulation, bioconcentration.**

- **biomarker**
1. parameter that can be used to identify a toxic effect in an individual organism and can be used in extrapolation between species.
2. indicator signalling an event or condition in a biological system or sample and giving a measure of exposure, effect, or susceptibility.

- **biomass**
  the amount of living matter in a given unit of the environment.

- **biomedical testing**
  testing of persons to find out whether a change in a body function might have occurred because of exposure to a hazardous substance.

- **bio-mineralisation**
  complete conversion of organic substances to inorganic derivatives by living organisms, especially micro-organisms.

- **bio-monitoring**
  the routine (e.g. monthly, quarterly) assessment of water, air, sediment or soil quality using the response(s) of living organisms as the measurement system.

- **biopsy**
  excision of a small piece of living tissue for microscopic or biochemical examination, usually performed to establish a diagnosis. See **autopsy.**

- **bio-remediation**
  the use of living organisms (e.g., bacteria) to clean up oil spills or remove other pollutants from soil, water, and wastewater.

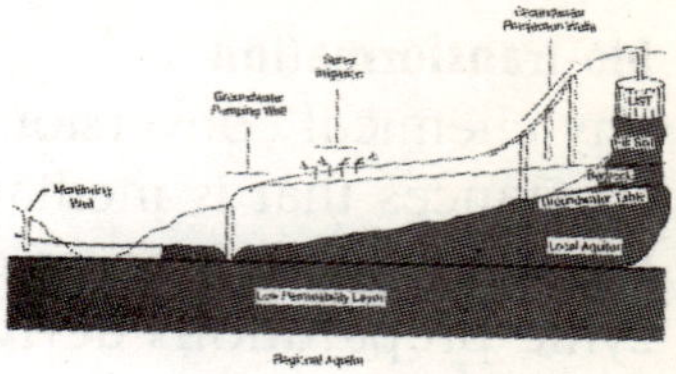

- **biosphere**
  portion of the planet earth which supports and includes life.

- **biostatic**
  arresting the growth or multiplication of living organisms.

- **biota**
  all living organisms as a totality.

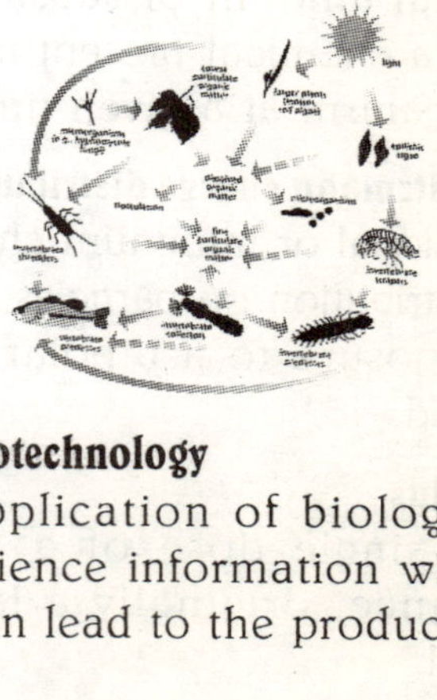

- **biotechnology**
  application of biological science information which can lead to the production

of commercial products by biologic processes. These include recombinant DNA techniques, development of monoclonal antibodies, as well as gene splicing.

- **bio-transformation**
  any chemical conversion of substances that is mediated by living organisms or enzyme preparations derived therefrom.

- **bipolar ion field**
  region in which ions of both polarities exist.
  see **exchange transfusion**.

- **bod5**
  the amount of dissolved oxygen consumed in five days by biological processes breaking down organic matter.

- **body burden**
  total amount of substance of a chemical present in an organism at a given time.

- **Boltzmann charge distribution**
  residual or minimum charge distribution on particles after exposure to a bipolar ion field.

- **bolus**
  1. single dose of a substance, originally a large pill.
  2. dose of a substance administered by a single rapid intravenous injection.
  3. concentrated mass of food ready to be swallowed.

- **boundary layer**
  region of flow near the bounding surface, where the flow is dominated by friction forces resulting in reduced flow velocity.

- **brady-**
  prefix meaning slow as in bradycardia or bradypnoea.

- **bradycardia**
  abnormal slowness of the heartbeat. See **tachycardia**.

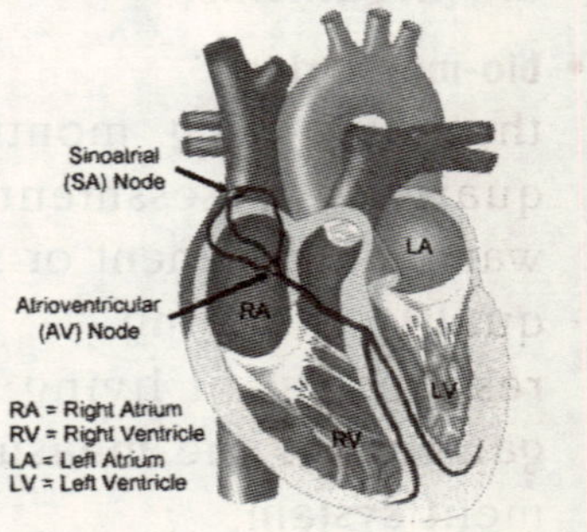

- **bradypnoea**
  abnormally slow breathing. See **tachypnoea**.

- **Brain Electrical Activity Mapping (BEAM)**
  a noninvasive diagnostic technique used to detect certain kinds of epilepsy,

head injuries, dyslexia, brain tumours, dementia, and other brain diseases. This procedure utilises a computer to convert data from brain electrical potentials into colored topographical maps of the brain which are then automatically compared with those of an appropriate control group to indicate the degree of deviation from the norm.

- **breathing zone**
  space within a radius of 0.5 m from a person's face.

- **breathing zone sample**
  a sample taken as close as possible to the point at which the subject inhales air. It represents a subject's inhaled air.

- **bronchoalveolar lavage fluid (BALF)**
  the fluid obtained by washing out the lungs for cytologic and enzymatic analyses.

- **bronchoconstriction**
  narrowing of the air passages through the bronchi of the lungs. See **bronchodilation**.

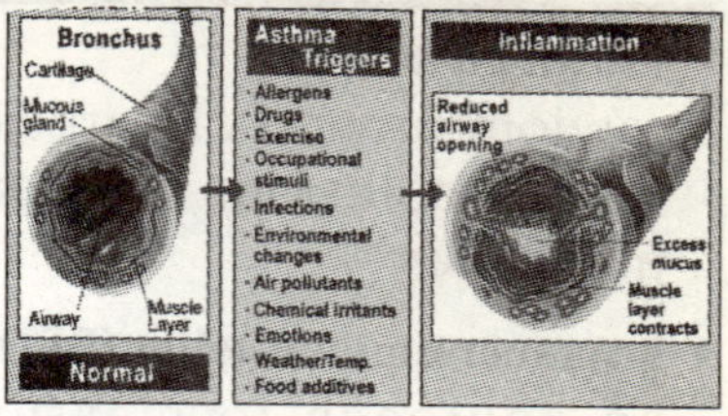

- **bronchodilation**
  expansion of the air passages through the bronchi of the lungs. see **bronchoconstriction**.

- **bronchoscopy**
  the direct visualisation of the trachea and bronchi through a rigid or flexible tube (bronchoscope). The procedure can be used to remove foreign bodies or to obtain a specimen for biopsy or other examination.

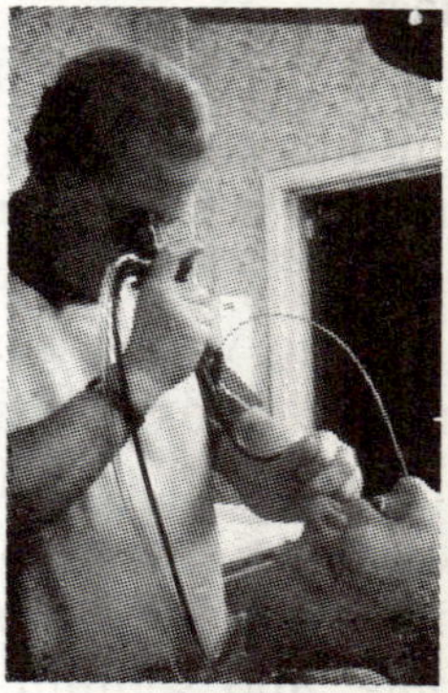

- **bronchospasm**
  intermittent violent contraction of the air passages of the lungs.

- **brownian motion**
random motion of particles due to collisions with gas molecules.

- **brunauer-emmett-teller (BET)**
an approach to measure the surface area of particles. The amount of material required to cover the particles with a single layer of molecules. It may be done with low-temperature adsorption of nitrogen or an inert gas from the vapour phase.

- **bubble meter**
a tube with a defined volume into which bubbles are injected to measure flow rate.

- **builder**
material which enhances or maintains the cleaning efficiency of a surfactant, in a detergent, principally by inactivating water hardness. Complex phosphates (especially sodium tripolyphospha i.e., pentasodium triphosphate), sodium carbonate, and sodium silicate are the builders most commonly used.

- **by-product**
materials, other than the intended product, generated as a result of an industrial process.

- **byssinosis**
pneumoconiosis caused by inhalation of dust and associated microbial contaminants and observed in cotton, flax, and hemp workers.

- **bystander exposure**
liability of members of the general public to come in contact with substances arising from operations or processes carried out by other individuals in their vicinity.

- **calcification**
process in which organic tissue becomes hardened by deposition of calcium salts within its substance.

- **calcium (Ca)**
an alkaline element found in most tissues, it is an abundant mineral in the body. It is essential for maintenance of bones and teeth,

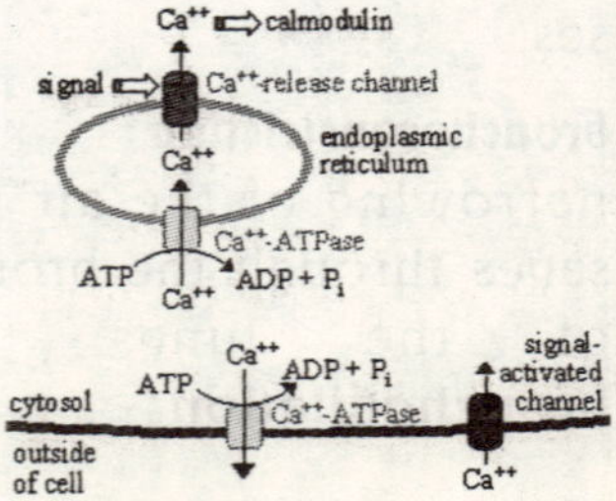

normal heart, muscle, and nerve functions, as well as normal blood clotting. Calcium levels can be measured in the blood or urine to help detect thyroid and parathyroid disorders, bone diseases, and certain cancers. It may also point to the presence of kidney stones.

- **calibration material**
  see **reference material**.

- **cancer**
  1. disease resulting from the development of a malignancy.
  2. any one of a group of diseases that occur when cells in the body become abnormal and grow or multiply out of control.

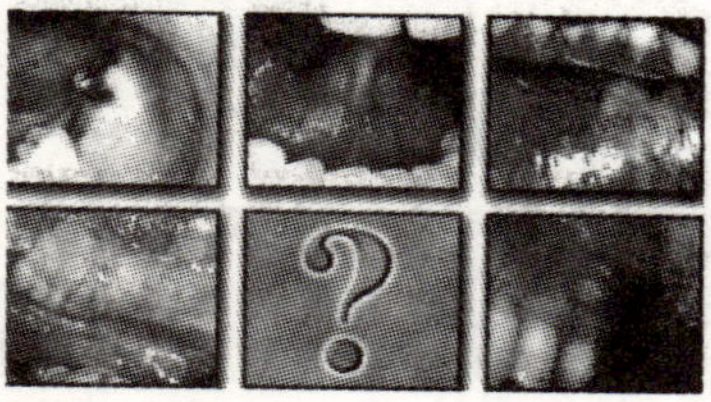

- **cancer risk**
  a theoretical risk for getting cancer if exposed to a substance every day for 70 years (a lifetime exposure) The true risk might be lower.

- **cannabinoids**
  the psychoactive chemical substances found in marijuana (Cannabis sativa). The presence of cannabinoids in urine or blood indicates recent marijuana use or close contact with it.

- **cap**
  an impermeable seal, usually composed of clay-type soil or a combination of clay soil and synthetic liner, which is placed over a landfill during closure. The cap serves to minimise leachate volume during biodegradation of the waste by keeping precipitation from percolating through the landfill. The cap also keeps odours down and animal scavengers from gathering.

- **capillary pore filter**
  a filter consisting of a solid membrane with an array of cylindrical holes of uniform size penetrating the membrane.

- **carbamazepine**
  a pain relieving and anticonvulsant drug often used to treat trigeminal

neuralgia (pain along the route of the facial nerve) and epilepsy (a neurological disorder characterised by convulsive seizures). Blood testing can determine carbamazepine levels in the body and aid in maintaining appropriate dosage.

- **carbohydrate**
  a group of organic compounds, including starches and sugars, that is a major source of body energy and is necessary for metabolism. Excessive intake is associated with tooth decay and obesity. Its concentration can be assayed in serum and urine.

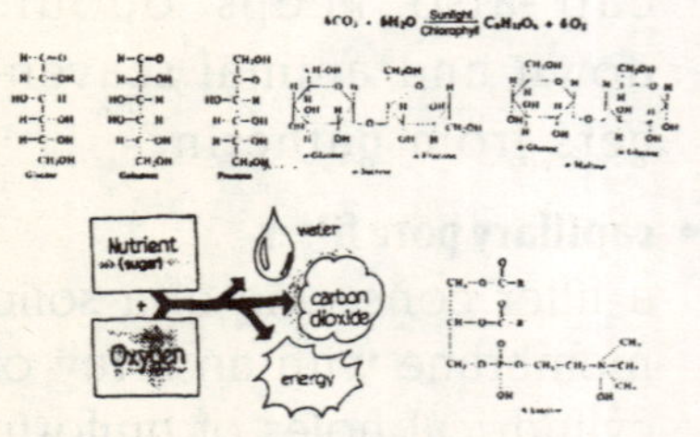

- **carbon dioxide (CO_2)**
  a colourless, odourless non-poisonous gas normally part of ambient air, a result of fossil fuel combustion.

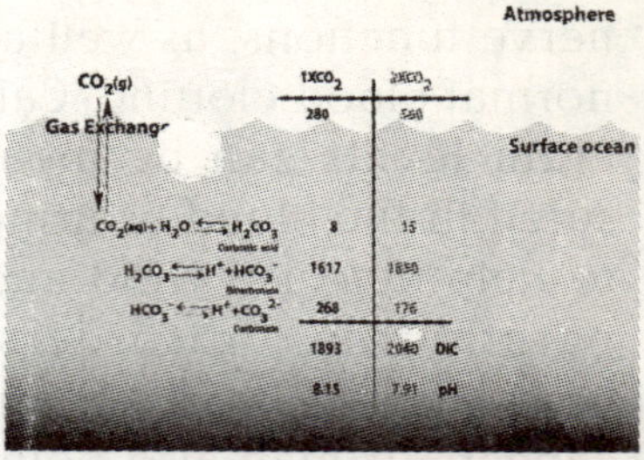

- **carbon monoxide (CO)**
  a colourless, odourless, poisonous gas produced by incomplete fossil fuel combustion.

- **carbonate hardness**
  that hardness in a water caused by bicarbonates and carbonates of calcium and magnesium. If alkalinity exceeds total hardness, all hardness is carbonate hardness, if hardness exceeds alkalinity, the carbonate hardness equals the alkalinity.

- **carboxyhaemoglobin**
  compound which is formed between carbon monoxide and haemoglobin in the blood of animals and which is incapable of transporting oxygen.

- **carcinogen**
  a carcinogen is any agent, chemical, physical or bio-

logical, that can act on living tissue in such a way as to cause a malignant neoplasm. More simply, a carcinogen is any substance which causes cancer

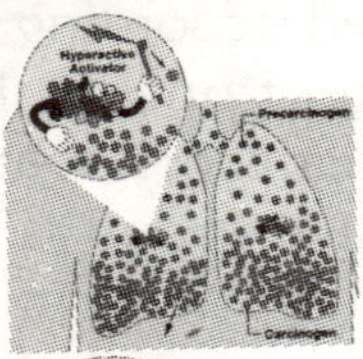

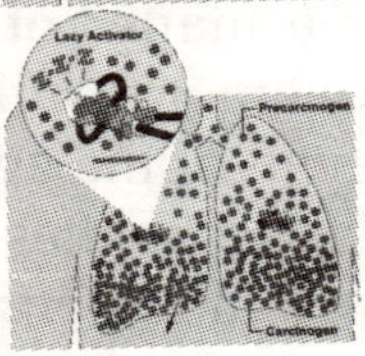

- **carcinogenesis**

carcinogenesis is the process which leads to development of cancer. Carcinogenesis may be a matter of induction by chemical, physical, or biological agents, of neoplasms that are usually not observed, an earlier induction of neoplasms that are usually observed, and/or the induction of more neoplasms than are usually found.

- **carcinogenic**

this is the adjective applied to anything which can cause cancer.

- **carcinogenicity**

a complex multistage process of abnormal cell growth and differentiation that can lead to cancer. Usually at least two stages and toxicants are involved: initiation in which a normal cell undergoes irreversible changes and promotion in which the initiated cells are stimulated to progress to cancer.

- **carcinogenicity test**

long term (chronic) test designed to detect any possible carcinogenic effect of a test substance.

- **carcinogenicity**

classification based on the weight of the evidence and not on potency as follows. Sufficient evidence. Causal relationship has been established between exposure to the agent and human cancer: a positive relationship has been observed between exposure to the agent and cancer in studies in which chance, bias and confounding could be ruled out with reasonable confidence.
Limited evidence. Positive association has been ob-

served between exposure to the agent and cancer for which a causal interpretation is considered to be credible, but chance, bias or confounding could not be ruled out with reasonable confidence.

Inadequate evidence. Available studies are of insufficient quality, consistency or statistical power to permit a conclusion regarding the presence or absence of a causal association.

Evidence suggesting lack of carcinogenicity. There are several adequate studies covering the full range of doses to which human beings are known to be exposed, which are mutually consistent in not showing a positive association between exposure to the agent and any studied cancer at any observed level of exposure. A conclusion of 'evidence suggesting lack of carcinogenicity' is inevitably limited to the cancer sites, circumstances and doses of exposure and length of observation covered by the available studies. In addition, the possibility of a very small risk at the levels of exposure studied can never be excluded.

Overall evaluation. Total body of evidence is taken into account, the agent is described according to the wording of one of the following categories, and the designated group is given. The categorisation of an agent is a matter of scientific judgement, reflecting the strength of the evidence derived from studies in humans and in experimental animals and from other relevant data.

Group 1 - The agent is carcinogenic to humans.

This category is used only when there is sufficient evidence of carcinogenicity in humans.

Group 2 - This category includes agents for which, at one extreme, the degree of evidence of carcinogenicity in humans is almost sufficient, as well as agents for which, at the other extreme, there are no human data but for which there is experimental evidence of carcinogenicity. Agents are assigned to either 2A (probably carcinogenic) or 2B (possibly carcinogenic) on the basis of epidemiological, experimental and other relevant data.

Group 2A - The agent is probably carcinogenic to humans. This category is used when there is limited evidence of carcinogenicity in humans and sufficient evidence of carcinogenicity in experimental animals. Exceptionally, an agent may be classified into this category solely on the basis of limited evidence of carcinogenicity in humans or of sufficient evidence of carcinogenicity in experimental animals strengthened by supporting evidence from other relevant data.

Group 2B - The agent is possibly carcinogenic to humans.

This category is generally used for agents for which there is limited evidence in humans in the absence of sufficient evidence in experimental animals. It may also be used when there is inadequate evidence of carcinogenicity in humans or when human data are nonexistent but there is sufficient evidence of carcinogenicity in experimental animals. In some instances, an agent for which there is inadequate evidence or no data in humans but limited evidence of carcinogenicity in experimental animals together with supporting evidence from other relevant data may be placed in this group.

Group 3 - The agent is not classifiable as to its carcinogenicity to humans. Agents are placed in this category when they do not fall into any other group.

Group 4 - The agent is probably not carcinogenic to humans. This category is used for agents for which there is evidence suggesting lack of carcinogenicity in humans together with evidence suggesting lack of carcinogenicity in experimental animals. In some circumstances, agents for which there is inadequate evidence or no data on carcinogenicity in humans but evidence suggesting lack of carcinogenicity in experimental animals, consistently and strongly supported by a broad range of other relevant data, may be classified in this group.

• **carcinoma**
malignant tumour of an epithelial cell. See **epithelioma.**

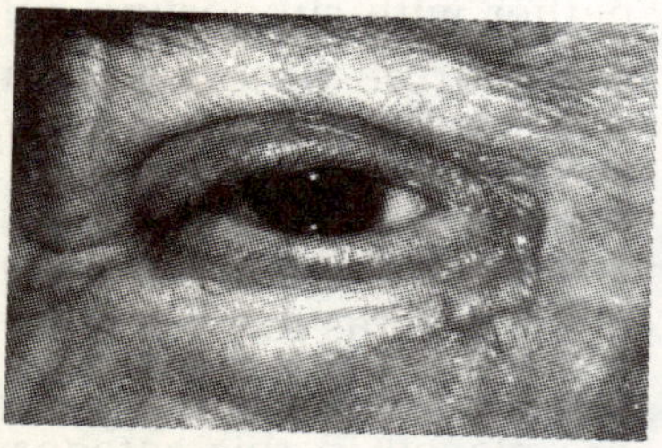

• **cardiotoxic**
chemically harmful to the cells of the heart.

• **carrying capacity**
1. in recreation, the amount of use a recreational area can sustain without deterioration of its quality.
2. in wildlife, the maximum number of animals an area can support during a given period of the year.

• **carry-over**
1. transfer in farming and agricultural processing of a component from one system such as soil or feed to another system such as a plant, animal or human being.
2. carry-over is expressed as the concentration of a component in the second system divided by the concentration in the first.
3. process in analytical studies by which materials are carried into a reaction mixture in which they do not belong.

• **cascade impactor**
a device that uses a series of impaction stages with decreasing particle cut size so that particles can be separated into relatively narrow intervals of aerodynamic diameter. Used for measuring the aerodynamic size distribution of an aerosol.

• **case control study**
a study which starts with the identification of persons with the disease (or other outcome variable) of interest, and a suitable control (comparison, reference) group of persons without the disease. The relationship of an attribute to the disease is examined by comparing the diseased

and non-diseased with regard to how frequently the attribute is present or, if quantitative, the levels of the attribute, in the two groups. See **case comparison study, case history study, case referent study, retrospective study.**

- **case study**
  a medical or epidemiologic evaluation of one person or a small group of people to gather information about specific health conditions and past exposures.

- **CAT scan (computer or computerised axial tomography)**
  see **computed tomography**.

- **catabolism**
  reactions involving the oxidation of organic substrates to provide chemically available energy (for example ATP) and to generate metabolic intermediates. Generally, process of breakdown of complex molecules into simpler ones, often providing biologically available energy. See **anabolism.**

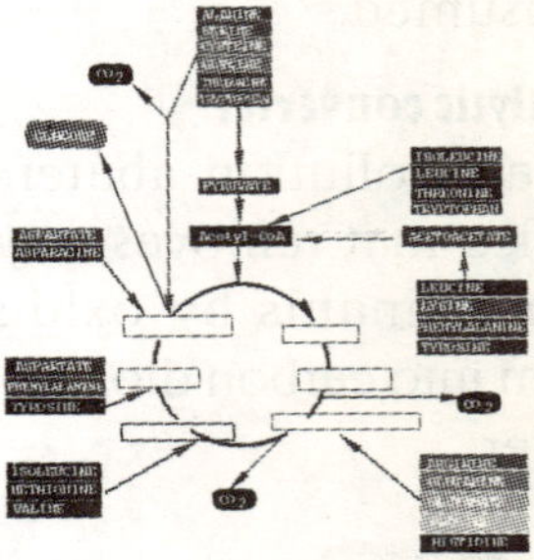

- **catalase**
  catalase is a haem-based enzyme which catalyses the breakdown of hydrogen peroxide into oxygen and water. It is found in all living cells, especially in the peroxisomes.

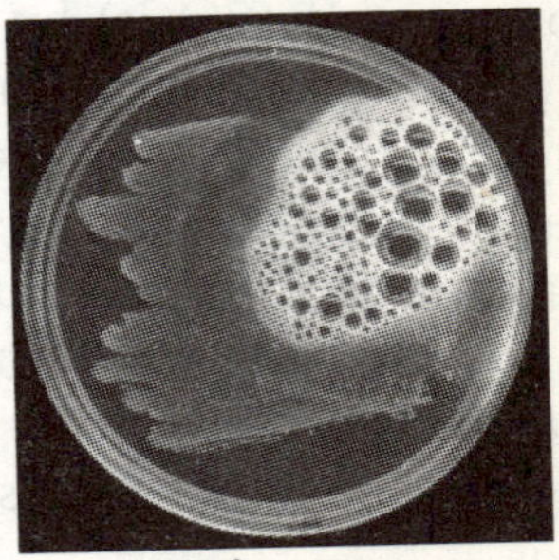

- **catalysis**
  addition of material (catalyst) that does not take a direct part in a chemical re-

action but increases the rate of the reaction.

- **catalyst**
  a substance that modifies (slows or quickens) a chemical reaction without being consumed.

- **catalytic converter**
  an air pollution abatement device that removes organic contaminants by oxidising them into carbon dioxide and water.

- **catalytic incinerator**
  a control device which oxidises volatile organic compounds (VOCs) by using a catalyst to promote the combustion process. Catalytic incinerators require lower temperatures than conventional thermal incinerators, with resultant fuel and cost savings.

- **catalyse**
  to accelerate the rate of a chemical reaction by a substance that is not permanently affected in the process.

- **catatonia**
  schizophrenia marked by excessive, and sometimes violent, motor activity and excitement, or by generalised inhibition.

- **cathartic**
  see **laxative**, **purgative**.

- **ceiling value (CV)**
  the ceiling value is the maximum permissible airborne concentration of a potentially toxic substance and is a concentration that should never be exceeded in the breathing zone.

- **cell line**
  defined unique population of cells obtained by culture from a primary implant through numerous generations.

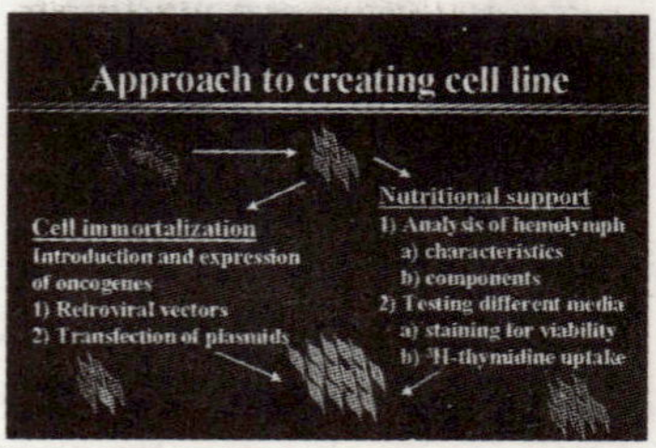

- **cell strain**
  cells having specific properties or markers derived from

a primary culture or cell line.

- **cell-mediated hypersensitivity**
state in which an individual reacts with allergic effects caused by the reaction of antigen-specific T-lymphocytes following exposure to a certain substance (allergen) after having been exposed previously to the same substance or chemical group. See **allergy, antigen, immunoglobulin E-mediated hypersensitivity**.

- **cell-mediated immunity**
immune response mediated by antigen-specific T-lymphocytes.

- **censored data**
sample observations for which the complete distribution is not known, for example, a cohort study in which some persons cannot be followed to the predetermined end of the study ('right-censored data') or environmental assay data in which some results are less than the sample detection limit ('left-censored data')

- **central nervous system**
the part of the nervous system that consists of the brain and the spinal cord.

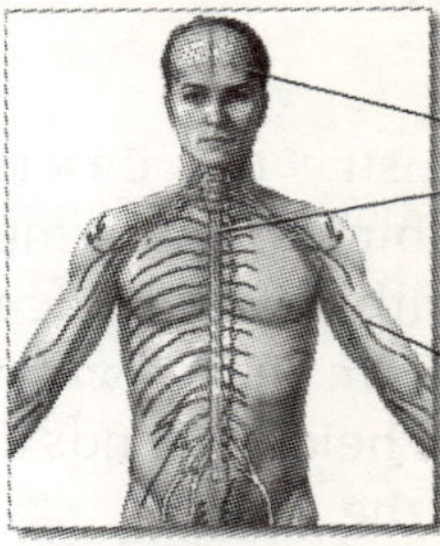

- **certified reference material**
reference material provided by a certifying body such as a National Standards Organisation or Metrological Laboratory or by an international body which confirms its purity and analytical values by technically valid procedures and provides a certificate detailing the relevant information. See **reference material**.

- **chain of custody**
sequence of responsibility for a substance from the manufacturer to the distributor, to the user, or to the person(s) ultimately responsible for waste disposal. This term is also used in controlled transmission of

samples from collection to analysis, especially of samples of materials used for medico-legal or forensic purposes.

- **chelation**
  in chemistry, this describes the combination of a metallic ion with heterocyclic ring structures in such a way that the ion is held by bonds from each of the rings.

- **chelation therapy**
  treatment with a chelating

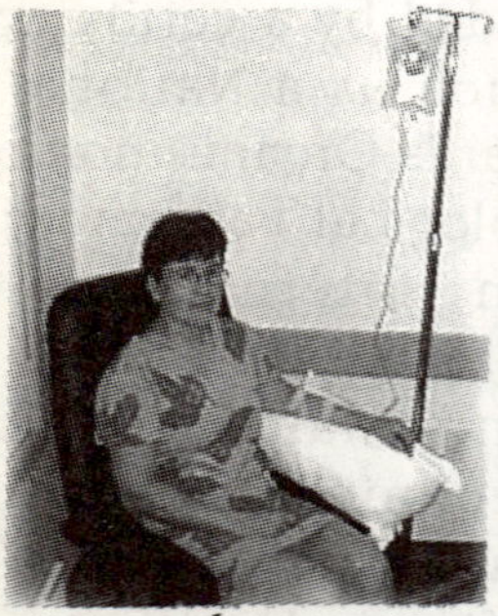

agent to enhance the elimination or reduce the toxicity of a metal ion.

- **chemical aetiologic agent**
  see **toxic substance**.

- **chemical conversion**
  change from one state or chemical structure to another. See **conversion**.

- **chemical etiologic agent**
  see **toxic substance**.

- **chemical oxygen demand (COD)**
  substance concentration of available oxygen (derived from a chemical oxidising agent) required to oxidise the organic (and inorganic) matter in waste water. See **biochemical oxygen demand**.

- **chemical precipitation**
  a chemical process in which a chemical in solution reacts with another chemical introduced to that solution to form a third substance which is partially or mainly insoluble and, therefore, appears as a solid.

- **chemical safety**
  practical certainty that there will be no exposure of organisms to toxic amounts of any substance or group of substances. This implies attaining an acceptably low risk of exposure to potentially toxic substances.

- **chemical species**
  set of chemically identical atomic or molecular structural units in a solid array or of chemically identical molecular entities that can explore the same set of mo-

lecular energy levels on the time scale of the experiment. For example, two conformational isomers may interconvert sufficiently slowly to be detectable by separate nuclear magnetic resonance spectra and hence be considered to be separate chemical species on a time scale governed by the radio-frequency of the spectrometer used. On the other hand, in a slow chemical reaction the same mixture of conformers may behave as a single chemical species, i.e., there is a virtually complete equilibrium population of the total set of molecular energy levels belonging to the two conformers. Except where the context requires otherwise, the term is taken to refer to a set of molecular entities containing isotopes in their natural abundance.

The wording of the definition given is intended to embrace both cases such as graphite, sodium chloride, or a surface oxide where the basic structural units are not capable of a separate existence as well as those cases where they are.

- **chemical toxicology**
see **toxicology**.

- **chemiluminescent**
exhibiting light as a result of a chemical reaction without the production of heat. Emitted from some bacteria, fungi, and fireflies.

- **chemistry**
in clinical testing, refers to the solutes dissolved in the plasma such as uric acid, etc.

- **chemobiokinetics**
see **toxicokinetics**.

- **chemophobia**
irrational fear of chemicals.

- **chemosis**
in medicine, this is the term given to any swelling around the eye caused by exposure to a chemical - the result of oedema of the conjunctiva.

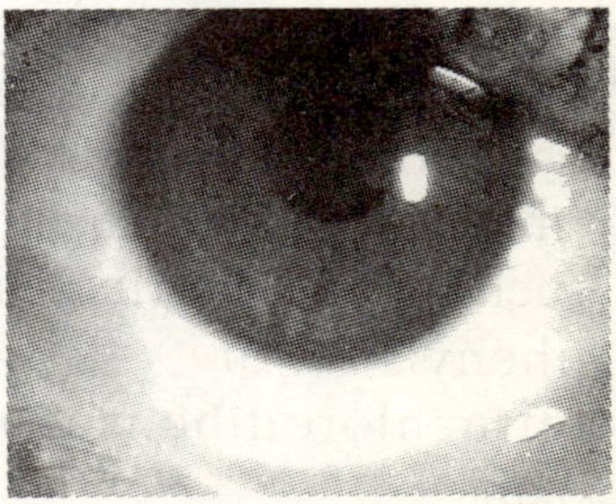

• **chemosteriliser**
substance used to sterilise mites, insects, rodents or other animals.

• **chlamydia trachomatis**
the causative micro-organism of one of the most prevalent sexually transmitted diseases, Chlamydia infection can lead to urethritis (inflammation of the urethra), pelvic inflammatory disease, ectopic pregnancy, and infertility. The infection may be asymptomatic, but patients may experience vaginal irritation or burning, discharge from penis or vagina, or pain. This microorganism is also the cause of serious eye infections. At one time testing for Chlamydia was somewhat difficult. Now, however, immunoassay can detect the antigens from swab specimens.

• **chloracne**
acne-like eruption cause by exposure to certain chlorinated organic substances such as polychlorinated biphenyls or 2,3,7,8-tetrachlorodibenzo-*p*-dioxin.

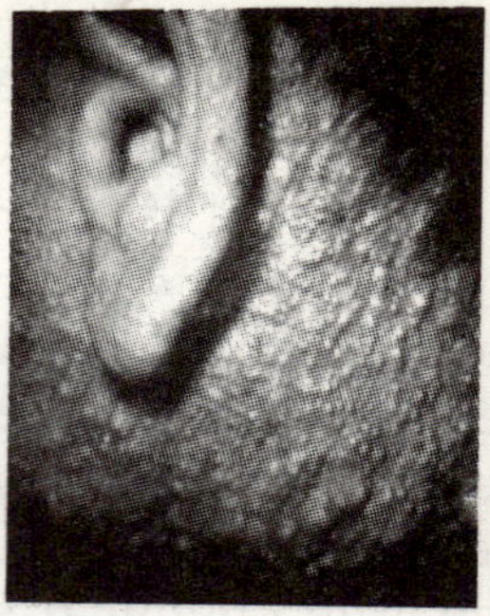

• **chloride**
a compound in which one of the elements is chlorine (sodium chloride — table salt — is the most common example). Found in the body in large amounts,it is an electrolyte, thus is involved

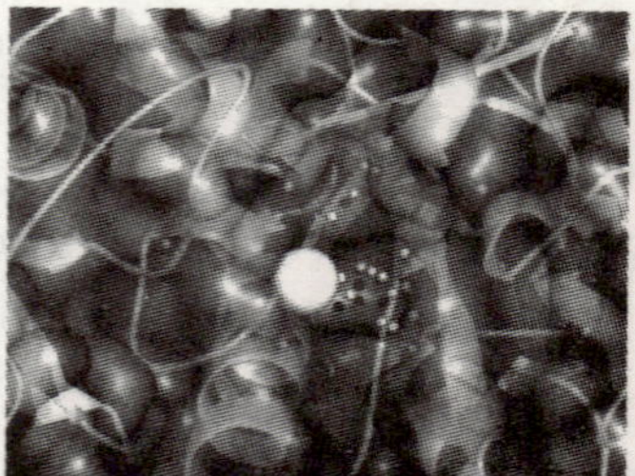

in water balance and acid-base balance. (Electrolytes affect movement of substances throughout the body and are crucial for normal function and metabolism.) To help determine electrolyte, fluid, kidney, and adrenal gland disorders, chloride levels usually are measured in blood, but also may be tested for in urine.

- **chlorination**
adding chlorine to water or wastewater, generally for the purpose of disinfection, but frequently for accomplishing other biological or chemical results. Chlorine also is used almost universally in manufacturing processes, particularly in the plastics industry.

- **chlorofluorocarbons (CFCs)**
a family of chemicals commonly used in air conditioners and refrigerators as coolants and also as solvents and aerosol propellants. CFCs drift into the upper atmosphere where their chlorine components destroy ozone. CFCs are thought to be a major cause of the ozone hole over Antarctica.

- **cholesterol**
a fat-soluble substance present in the blood and in animal fats that facilitates absorption and transport of fatty acids, and is necessary in hormone production. High levels of blood cholesterol have been linked to atherosclerosis and heart disease. See **high and low density lipoprotein cholesterols**.

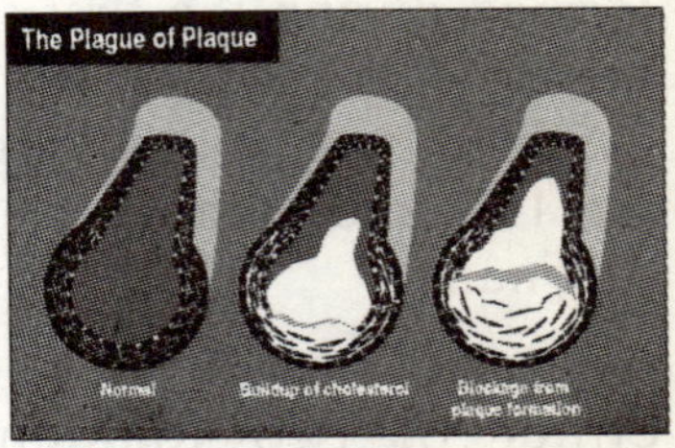

- **cholinesterase and pseudocholinesterase inhibitors**
these are substances which inhibit the enzyme cholinesterase and thus enhance and subsequently prevent transmission of nerve impulses from one nerve cell to another or to a muscle.

- **cholinesterase inhibitor**
substance which inhibits the action of acetylcholinesterase (EC 3.1.1.7) and related enzymes which catalyse the hydrolysis of choline esters. Such a substance causes hyperactivity in parasympathetic nerves.

- **cholinomimetic**
see **parasympathomimetic.**

- **chorionic villi sampling**
a means of testing the foetus for various abnormali-

ties, such as congenital defects, blood disorders, and inborn errors of metabolism. This test can be performed during the first trimester of pregnancy, but is easier to carry out between the 14th and 16th weeks of pregnancy.

- **chromatid**
either of two filaments joined at the centromere which make up a chromosome.

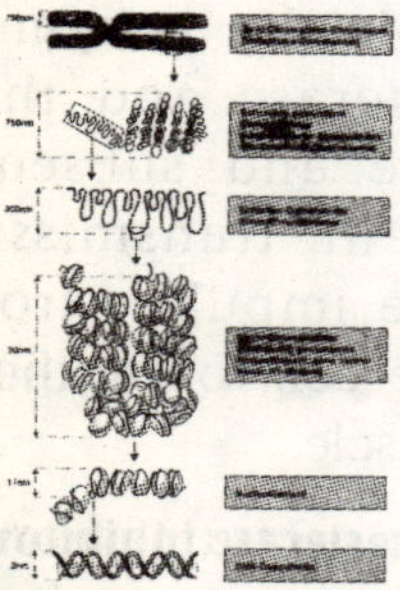

- **chromatin**
stainable complex of DNA and proteins present in the nucleus of a eukaryotic cell. See **eukaryote**.

- **chromosomal aberration**
abnormality of chromosome number or structure.

- **chromosome**
1. self-replicating structure consisting of DNA complexed with various proteins and involved in the storage and transmission of genetic information. The physical structure that contains the genes. See **chromatid**.

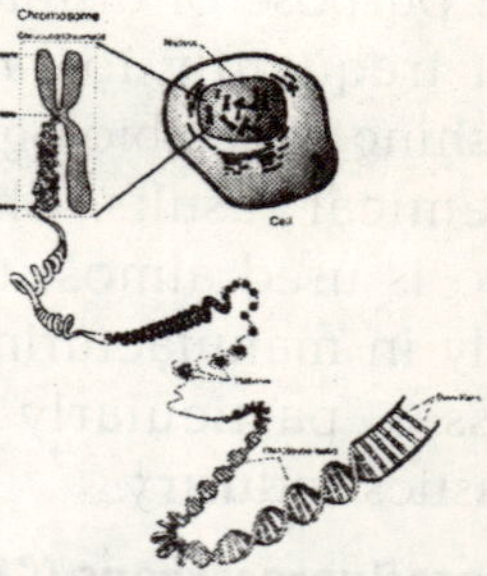

2. this is a structure in the nucleus of a cell composed of deoxyribonucleic acid (DNA) and protein; the chromosome forms the basis of heredity and carries genetic information in DNA in the form of a sequence of nitrogenous bases.

- **chronic**
occurring over a long time [compare with **acute**].

- **chronic carrier**
someone who has been infected by a pathogenic organism and has no symptoms of the illness, but has the ability to infect others.

- **chronic effects**
consequences which develop slowly and have a

long-lasting course (often but not always irreversible).See **acute effect.**

- **chronic exposure**
continued exposures occurring over an extended period of time, or a significant fraction of the test species' or of the group of individuals', or of the population's life-time. See **acute exposure**.

- **chronic hepatitis B**
long-term infection with the hepatitis B virus. (Hepatitis B also has been called serum hepatitis.) Such infection may lead to liver cell destruction and cirrhosis.

- **chronic test**
a comparative study in which organisms, that are subjected to different treatments, are observed for a long period or a substantial portion of their life span. Acute tests often utilise mortality as the only measured effect while chronic tests usually include additional measures of effect such as growth or reproduction . Note Cost-effective short-term chronic tests have been designed which involve exposure of sensitive life stages (e.g. larval fish, juvenile invertebrates) for a relatively brief period (e.g. seven days).

- **chronic toxicity**
1. adverse effects following chronic exposure.
2. effects which persist over a long period of time whether or not they occur immediately upon exposure or are delayed. See **acute toxicity**.

- **chronic toxicity test**
study in which organisms are observed during the greater part of the life span and in which exposure to the test agent takes place over the whole observation time or a substantial part thereof. See **acute toxicity test**, **long term. test.**

- **chronic value (ChrV)**
the geometric mean of the NOEC and LOEC. For example NOEC = 25 mg/l, LOEC = 50 mg/l ChrV = antilog[(log 25 + log 50)/2] = 35.3 mg/l.

• **chronotoxicology**
study of the influence of biological rhythms on the toxicity of substances.
circulation of substances in the environment movement of xenobiotic substances in the environment with air flow, river current, soil, water, etc. See **biological cycle.**

• **cirrhosis**
1. liver disease defined by histological examination and characterised by increased fibrous tissue, abnormal physiological changes such as loss of functional liver cells, and increased resistance to blood flow through the liver.

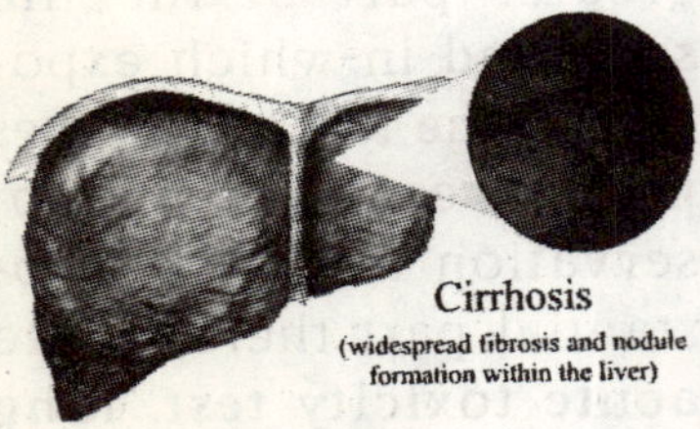

Cirrhosis
(widespread fibrosis and nodule formation within the liver)

2. interstitial fibrosis of an organ.

• **clastogen**
agent causing chromosome breakage and/or consequent gain, loss or rearrangement of pieces of chromosomes.

• **clastogenesis**
occurrence of chromosomal breaks and/or consequent gain, loss or rearrangement of pieces of chromosomes.

• **clastogenic**
this is the adjective applied to any substance or process causing chromosomal breaks.

• **clearance**
1. volume of blood or plasma or mass of an organ effectively cleared of a substance by elimination (metabolism and excretion) in a given time interval. Clearance is expressed in units of volume or mass per unit of time. Total clearance for a component is the sum of the clearances of each eliminating organ or tissue for that component.
2. in pulmonary toxicology, clearance refers specifically to removal of any inhaled substance which deposits on the lining surface of the lung. Lung clearance is expressed in volume or mass of lung cleared per unit time.
3. in renal toxicology, clearance refers to the quantifi-

cation of the removal of a substance by the kidneys by the processes of filtration and secretion. Clearance is calculated by relating the rate of renal excretion to the plasma concentration. See **elimination**.

- **climate change**

this term is commonly used interchangeably with 'global warming' and 'the greenhouse effect', but is a more descriptive term. Climate change refers to the build-up of man-made gases in the atmosphere that trap the suns heat, causing changes in weather patterns on a global scale. The effects include changes in rainfall patterns, sea level rise, potential droughts, habitat loss, and heat stress. The greenhouse gases of most concern are carbon dioxide, methane, and nitrous oxides. If these gases in our atmosphere double, the earth could warm up by 1.5 to 4.5 degrees by the year 2050, with changes in global precipitation having the greatest consequences.

- **clone**

1. population of genetically identical cells or organisms having a common ancestor.
2. to produce such a population.
3. recombinant DNA molecules all carrying the same inserted sequence.

- **clonic**

pertaining to alternate muscular contraction and relaxation in rapid succession. See **tonic**.

- **closure**

the procedure an operator must go through when a landfill reaches the legal capacity for solid waste. No more waste can be accepted and a cap usually is placed over the site. The cap is then planted with grasses and other ground covers. Post-closure care includes monitoring ground water, landfill gases, and leachate collection systems, sometimes for as long as 30 years.

- **cluster investigation**

a review of an unusual number, real or perceived, of health events (for example, reports of cancer) grouped together in time and location. Cluster investigations are designed to confirm case reports, determine whether they represent an unusual disease occurrence, and explore possible causes and contributing environmental factors.

- **cluster sampling**

1. a method of sampling in which the population is divided into aggregates (or clusters) of items bound together in a certain manner. A sample of these clusters is taken at random and all the items which constitute them are included in the sample.

2. a sampling method in which each unit selected is a group of persons (all persons in a city block, a family, etc.) rather than an individual.

- **coagulation**

a chemical reaction in which polyvalent ions neutralise the repulsive charges surrounding colloidal particles. The clumping together of solids to make them settle out of solution faster. Coagulation of solids is brought about with the use of certain chemicals, such as lime, alum or polymers.

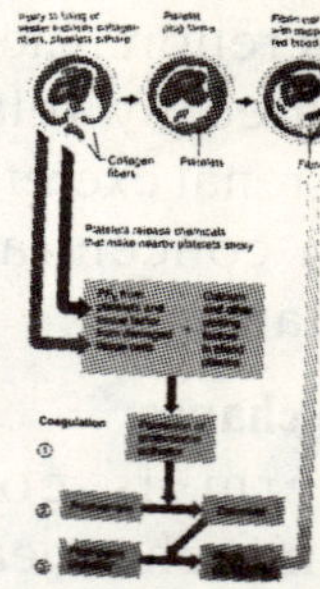

- **coagulation and flocculation**

process which follow sequentially and are distinguished primarily by the types of chemicals used for their initiation and the size of the particles developed. Coagulation is the conversion of finely dispersed colloids into small floc upon the addition of electrolytes such as inorganic acids, bases and salts.

- **coagulation chemicals**

hydrolysable divalent and trivalent metallic ions of aluminium (Al^{3+}), magnesium (Mg^{2+}) and iron (Fe^{3+}) salts. They include alum (aluminium sulphate: $Al_2(SO_4)_3.14\ H_2O$), quicklime (calcium oxide: CaO), sulphuric acid (H_2SO_4), anhydrous ferric chloride

($FeCl_3$). Lime and acid affect only the solution pH which in turn causes coagulant precipitation, such as that of magnesium.

- **coalescence**
the gathering together of coagulated colloidal liquid particles into a single continuous phase.

- **cocaine**
a substance used as a local anaesthetic, also used as a recreational inhalant, that can lead to psychological dependence with prolonged use. Its use can cause damage to mucous membranes and cardiovascular disorders. Cocaine metabolites can be detected in urine and may indicate recent use of the drug.

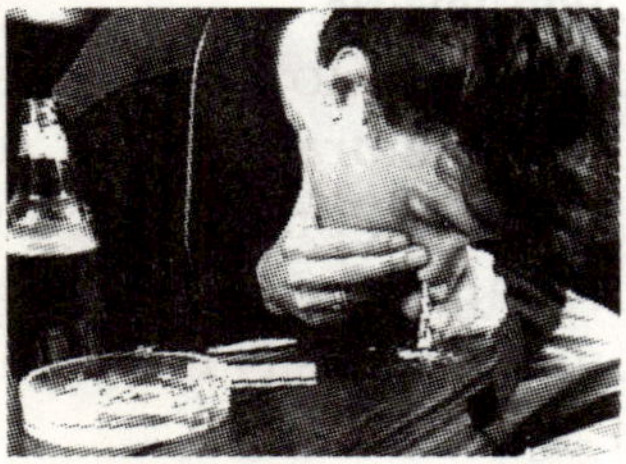

- **cocarcinogen**
chemical, physical or biological factor which intensifies the effect of a carcinogen.

- **coefficient of variation (CV)**
a standard statistical measure of the relative variation of a distribution or set of data, defined as the standard deviation divided by the mean. It is also called the relative standard deviation (RSD). The CV can be used as a measure of the precision within and between laboratories, or among replicates for each treatment concentration

- **cohort**
component of the population born during a particular period and identified by period of birth so that its characteristics (such as causes of death and numbers still living) can be ascertained as it enters successive time and age periods. The term 'cohort' has broadened to describe any designated group of persons followed or traced over a period of time, as in the term cohort study (prospective study).

- **cohort analysis**
tabulation and analysis of morbidity or mortality rates in relationship to the

ages of a specific group of people (cohort), identified by their birth period, and followed as they pass through different ages during part or all of their life span. In certain circumstances such as studies of migrant populations, cohort analysis may be performed according to duration of residence in a country rather than year of birth, in order to relate health or mortality experience to duration of exposure.

- **colonoscopy**

visualisation of the lining of the anus, rectum and colon through a rigid proctosigmoidoscope or a flexible fibre optic endoscope (types

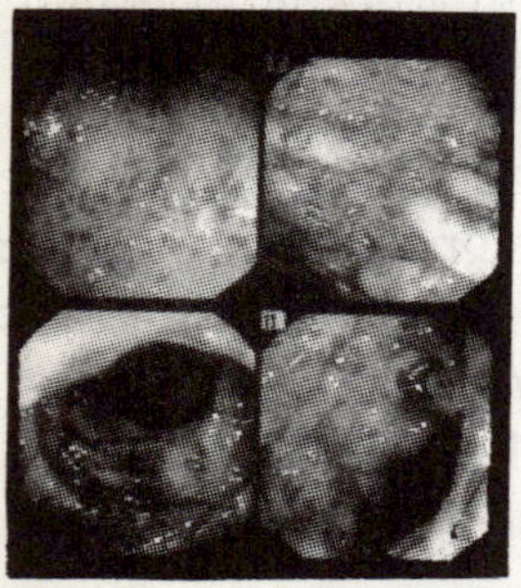

of viewing tubes). This procedure allows diagnosis of tumours and inflammatory diseases.

- **colorectal cancer**

a malignant disease of the colon and/or rectum which often begins as a polyp. The first indication of colorectal cancer is usually hidden, or occult, blood in the stool. Symptoms may include changes in bowel habits (diarrhoea, constipation) or in consistency or colour of the stool. A stool specimen blood test is commonly performed to screen for colorectal cancer.

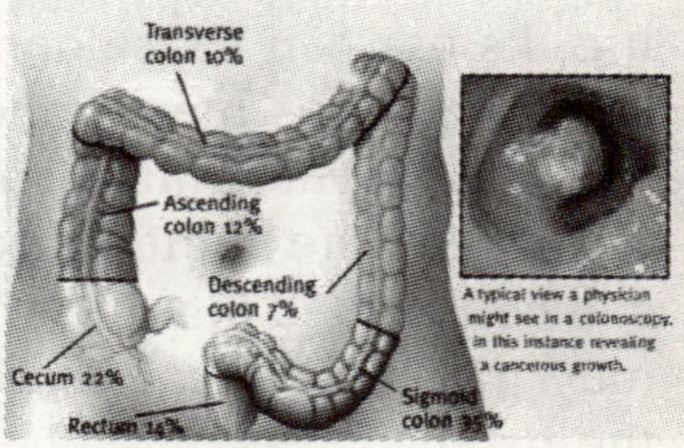

- **colourimetric**

pertaining to measurement of concentration of a solution based on its absorption or transmission of light, or on the intensity of colour in a liquid.

- **combination testing**

a series of independent tests done on single specimens to arrive at a medical decision.

- **combined effect of poisons**
simultaneous or successive effect of two or more poisons on the organism by the same route of exposure. See **additive effect, antagonism, independent effects of poisons, potentiation, summation, synergism**.

- **combined sewers**
a sewer system that carries both sewage and storm-water runoff. Normally, its entire flow goes to a waste treatment plant, but during a heavy storm, the storm water volume may be so great as to cause overflows. When this happens, untreated mixtures of storm water and sewage may flow into receiving waters. Storm-water runoff may also carry toxic chemicals from industrial areas or streets into the sewer system.

- **cometabolism**
process by which a normally non-biodegradable substance is biodegraded only in the presence of an additional carbon source. See **analogue metabolism**.

- **commercial waste**
all solid waste from businesses. This category includes, but is not limited to, solid waste originating in stores, markets, office buildings, restaurants, shopping centres, and theatres.

- **commercial waste management facility**
a treatment, storage, disposal, or transfer facility that accepts wastes from a variety of sources for profit. A commercial facility manages a broader spectrum of wastes than a private facility, which normally manages a limited volume or type of waste.

- **comparison group**
see **control group**.

- **comparison value (CV)**
calculated concentration of a substance in air, water, food, or soil that is unlikely to cause harmful (adverse)

health effects in exposed people. The CV is used as a screening level during the public health assessment process. Substances found in amounts greater than their CVs might be selected for further evaluation in the public health assessment process.

- **compartment**
  part of the body considered as an independent system for purposes of assessment of distribution and clearance of a substance. The body is composed of a large number of organs, tissues, cells, cell organelles and fluids, any one of which could be referred to as a compartment. In kinetic considerations, a compartment often refers collectively to the organs, tissues, cells, and fluids for which the rates of uptake and subsequent distribution and elimination are sufficiently similar to preclude kinetic resolution.

- **compensation**
  adaptation of an organism to changing conditions of the environment (especially chemical) is accompanied by the emergence of stresses in biochemical systems which exceed the limits of normal (homeostatic) mechanisms. Compensation is a temporary concealed pathology which later on can be manifested in the form of explicit pathological changes (decompensation). See **pseudo-adaptation, acclimatisation, adaptation**.

- **competent bacteria**
  culture of bacteria (or yeast) treated in such a way that their ability to take up DNA molecules without transduction or conjugation has been enhanced.

- **competitive-binding assay**
  an analytic procedure in which an unlabelled and a labelled analyte compete for sites to bind to a specific protein.

- **complement**
  a group of serum proteins that destroy sensitised cells or bacteria.

- **complete blood count (CBC)**
  the determination of the blood cells in a definite vol-

ume of blood that includes separate counts for red and white blood cells.

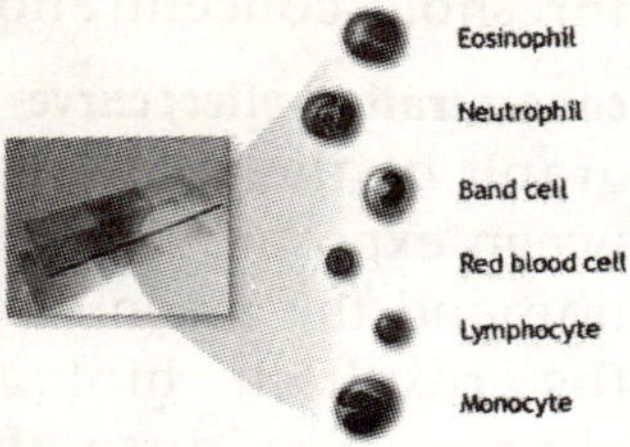

- **complete mineralisation**
  complete breakdown of a complex organic compound to carbon dioxide, water, oxides and oxidative inorganic products such as nitrate or sulphate.
- **completed exposure pathway**
  see **exposure pathway**.
- **composite sample**
  a sample which is made by combining samples from two or more locations. The sample can be of water, soil or another medium.
- **compost**
  decomposed organic material that is produced when bacteria in soil break down garbage and biodegradable trash, making organic fertiliser. Making compost requires turning and mixing and exposing the materials to air. Gardeners and farmers use compost for soil enrichment.

- **composting**
  a controlled microbial degradation of organic waste yield an environmentally sound, nuisance-free product of potential value as a soil conditioner.
- **compound**
  a substance composed of two or more elements (see **element**) whose composition is constant. For example, table salt (sodium chloride NaCl) is a compound.
- **comprehensive effect of poisons**
  simultaneous or successive effect made on an organism by poisons entering from different media, from air, from water, from food or through the skin.
- **computer tomography (CT)**
  a radiographic scanning procedure to obtain cross-sectional images. The X-ray

source rotates about the patient, generating X-rays which pass through the individual and are monitored as they exit, by detectors that are linked to a computer. The computer constructs three-dimensional images for examination and analysis. CT has revolutionised the diagnosis of intracranial, intrathoracic, and intraabdominal tumours, as well as other space-occupying lesions. CAT scan (computer axial tomography), CTAT (computerised transaxial tomography), CRT (computerised reconstruction tomography), and DAT (digital axial tomography) are diagnostic radiographic tools that are now collectively called computed tomography.

• **concentration**
the amount of a substance present in a certain amount of soil, water, air, food, blood, hair, urine, breath, or any other media. See **absolute lethal concentration, lethal concentration, maximum tolerable concentration, median effective concentration, median lethal concentration, median narcotic concentration, minimum lethal concentration, threshold concentration**.

• **concentration-effect curve**
graph of the relation between exposure concentration and the magnitude of the resultant biological change. See **dose-effect curve, exposure effect curve, dose-effect relationship**.

• **concentration-response curve**
graph of the relation between exposure concentration and the proportion of individuals in a population responding with a quantal effect. See **dose-response curve, response**.

• **concentration-response relationship**
association between exposure concentration and the incidence of a defined biological effect in an exposed population. See **dose-response relationship, response**.

• **concordance**
pairs or groups of individuals of identical phenotype. For example, in twin studies, a condition in which both

twins exhibit or fail to exhibit a trait under investigation. See **phenotype**.

- **concurrent study**
  see **cohort study**.

- **concurrent validity**
  measurement and its criterion refer to the same point in time. An example would be a visual inspection of a wound for evidence of infection validated against bacteriological examination of a specimen taken at the same time.

- **condensation**
  a physical process with more vapour molecules arriving at particle's surface than leaving the surface, resulting in a net growth of the particle.

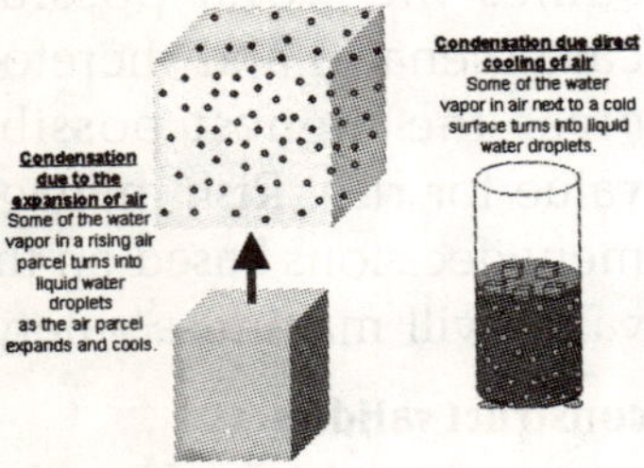

- **condensation nuclei counter**
  a device in which submicrometre-sized particles are grown by vapour supersaturation to a larger size and are detected by light scattering.

- **confidence limits (or intervals)**
  these are probability limits, based on the data set and statistical model employed, that the 'true value' lies within the limits specified. Typically, limits are expressed at the 95% or 99% probability levels.

- **confounding variable**
  changing factor that can cause or prevent the outcome of interest, is not an intermediate variable, and is not associated with the factor under investigation. Such a variable must be controlled in order to obtain an undistorted estimate of the effect of the study factor on risk. See **confounder**.

- **confounding**
  1. situation in which the effects of two processes are not distinguishable from one another.
  2. the distortion of the apparent effect of an exposure on risk brought about by the association of other factors which can influence the outcome

3. relationship between the effects of two or more causal factors as observed in a set of data, such that it is not logically possible to separate the contribution which any single causal factor has made to an effect.
4. Situation in which a measure of the effect of an exposure on risk is distorted because of the association of exposure with other factor(s) which influence the outcome under study.

- **congener**
substance which by structure, function or origin is similar to another.

- **conjugate**
1. derivative of a substance formed by its combination with compounds such as acetic acid, glucuronic acid, glutathione, glycine, sulfuric acid etc. See **phase 2 reaction**.
2. material produced by attaching two or more substances together, for example - conjugates of antibody with fluorochromes, radio-isotopes or enzymes.

- **conjunctiva**
mucous membrane which covers the eyeball and lines the under-surface of the eyelid.

- **conjunctivitis**
inflammation of the conjunctiva.

- **conservation**
preserving and renewing natural resources to assure their highest economic or social benefit over the longest period of time. Clean rivers and lakes, wilderness areas, a diverse wildlife population, healthy soil, and clean air are natural resources worth conserving for future generations.

- **conservative assessment of risk**
assessment of risk which assumes the worst possible case scenario and therefore gives the highest possible value for risk. Risk management decisions based on this value will maximise safety.

- **construct validity**
extent to which the measurement corresponds to theoretical concepts (constructs) concerning the phenomenon under study. For example, if on theoreti-

cal grounds, the phenomenon should change with age, a measurement with construct validity would reflect such a change.

- **construction and demolition waste**
waste building materials, dredging materials, tree stumps, and rubble resulting from construction, re-modelling, repair, and demolition operations on houses, commercial buildings and other structures, and pavements. May contain lead, asbestos, or other hazardous materials.

- **contact dermatitis**
inflammatory condition of the skin resulting from dermal exposure to an allergen (sensitiser) or an irritating (corrosive) substance.

- **containment**
process by which possible release, discharge or spill of a toxic substance during normal use or after an accident is prevented by appropriate action.

- **contaminant**
1. minor impurity present in a substance.
2. extraneous material inadvertently added to a sample prior to or during chemical or biological analysis.
3. in some contexts, as in relation to gas cleaning equipment, used as a synonym for 'pollutant', especially on a small scale.
4. unintended component in food that may pose a hazard to the consumer.

- **contamination**
intrusion of undesirable elements. The addition of foreign matter to a substance which reduces the value of the substance, or interferes with its intended use.

- **content validity**
extent to which the measurement incorporates the domain of the phenomenon under study. For example, a measurement of functional

health status should embrace activities of daily living, occupational, family, and social functioning, etc.

- **continuous discharge**
  a permitted release of pollutants into the environment that occurs without interruption, except for infrequent shutdowns for maintenance, process changes, etc.

- **continuous treatment**
  treatment of waste streams operating without interruption as opposed to batch treatment, sometimes referred to as flow-through treatment.

- **contraindication**
  any condition which renders some particular line of treatment improper or undesirable.

- **control chart**
  a cumulative summary chart of results from QA tests with reference materials (e.g. reference toxicants). The results of a given QA test are compared to the control chart mean value and acceptance limits (typically 95% confidence limits, i.e. mean + 2 standard deviations) or warning limits (typically 99% confidence limits, i.e. mean + 3 standard deviations).

- **control group**
  selected group, identified as a rule before a study is done, which comprises humans, animals, or other species who do not have the disease, intervention, procedure or whatever is being studied, but in all other respects is as nearly identical to the test group as possible. See **comparison group**.

- **control limit**
  a regulatory value applied to the airborne concentration in the workplace of a potentially toxic substance which is judged to be 'reasonably practicable' for the whole spectrum of work activities and which must not normally be exceeded.

- **control, matched**
  control (individual or group or case) selected to be similar to a study individual or group, or case, in specific characteristics: some commonly used matching variables are age, sex, race and socio-economic status.

• **control, negative**
measures taken to insure that a test, its components, or the environment do not cause undesired effects, or produce incorrect test results. In toxicity tests a negative control typically consists of clean water or sediment that is essentially free of contaminants or the test material. The response of organisms in the negative control is used to determine test acceptability and as a baseline for gauging adverse effects among animals exposed to treatments.

• **control, positive**
measures taken to insure that a test and/or its components are working properly and producing correct or expected results from positive test subjects In toxicity tests positive controls typically consist of reference toxicants.

• **controlled reaction**
a chemical reaction at temperature and pressure conditions that are maintained within safe limits to produce a desired product.

• **conventional filtration treatment**
means a series of processes including coagulation, flocculation, sedimentation, and filtration resulting in substantial particulate removal.

• **conversion**
see **chemical conversion, bio-transformation.**

• **core window**
the period during which a person infected by the hepatitis B virus may infect others, even though blood tests show neither hepatitis B surface antigen nor antibodies to hepatitis core antigen.

• **corrosion of tissue**
this is the destruction of tissue by a substance on direct contact.

• **corrosive**
causing a surface-destructive effect on contact. In toxicology, this normally means causing visible destruction of the skin, eyes, or the lining of the respiratory tract or the gastrointestinal tract.

- **corrosive of tissue**
the descriptor applied to any substance which destroys tissues on direct contact.

- **count mean diameter**
mean of the diameters of all particles in a population.

- **count median diameter**
calculated diameter in a population of particles in a gas or liquid phase above which there are as many particles with larger diameters as there are particles below it with smaller diameters.

- **crackles**
see **crepitations**.

- **c-reactive protein (CRP)**
a protein that appears, usually within 24 hours, in the blood during the acute stage of inflammatory disorders such as rheumatic fever, or after a myocardial infarction (heart attack). The serum level of CRP is a sensitive indicator and monitor of rheumatic activity.

- **creatine kinase (CK)**
an isoenzyme (an enzyme that appears in multiple forms) found in high concentrations in heart and skeletal muscles, and, in smaller amounts, in brain tissue. Each form of CK is predominant in specific tissues. Determining blood levels of the isoenzyme will aid in detection of myocardial infarction and muscle diseases such as muscular dystrophy.

- **creatinine**
found in muscle and blood and excreted in urine, creatinine is the end product of muscle (creatine) metabolism. By ascertaining levels of creatinine in blood or urine or in both, presence or severity of kidney malfunction can be indicated.

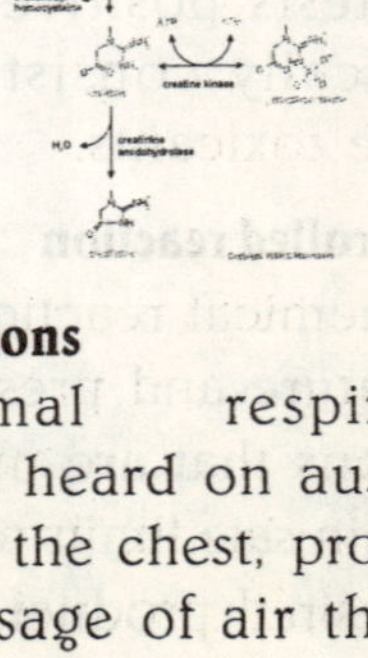

- **crepitations**
abnormal respiratory sounds heard on auscultation of the chest, produced by passage of air through

passages which contain secretion or exudate or which are constricted by spasm or a thickening of their walls, more usually referred to as crepitations or rhonchi (auscultation is the process of listening for sounds within the body by ear unassisted or using a stethoscope). See **crackles**.

- **criteria**
descriptive factors taken into account by the Environment Agency in setting standards for pollutants. For example, water quality criteria describe the concentration of pollutants that most fish can be exposed to for an hour without showing acute effects.

- **criteria continuous concentration (CCC)**
water quality criteria for chronic exposure.

- **criteria maximum concentration (CMC)**
water quality criteria for acute exposure.

- **criterion**
validated set of data used as a basis for judgement.

- **criterion validity**
extent to which the measurement correlates with an external criterion of the phenomenon under study.

- **critical concentration (for a cell or organ)**
concentration of a potentially toxic substance at which undesirable (or adverse) functional changes, reversible or irreversible, occur in the cell or organ.

- **critical effect**
for deterministic effects, the first adverse effect which appears when the threshold (critical) concentration or dose is reached in the critical organ. Adverse effects, such as cancer, with no defined threshold concentration are often regarded as critical. Decision on whether an effect is critical is a matter of expert judgment.

- **critical end-point**
toxic effect used as the basis for a reference dose. See **reference dose**.

- **critical group**
part of a target population most in need of protection

because it is most susceptible to a given toxicant.

- **critical organ**
1. in toxicology, organ which first attains the critical concentration (of a potentially toxic substance) under specified circumstances of exposure and for a given population.
2. in radiation biology, organ the damage of which (by radiation) results in the greatest injury to the individual (or his/her descendants). The injury may result from inherent radiosensitivity or indispensability of the organ, or from high dose, or from a combination of all three.

- **critical organ concentration (of a substance)**
mean concentration in the critical organ at the time the most sensitive type of cell reaches the critical concentration. See **critical concentration, critical organ**.

- **critical orifice**
a very small hole through which there will be constant airflow with even higher head pressure after a sufficient pressure drop across the orifice causes sonic flow.

- **critical period (of development)**
stage of development of an organism that is of particular importance in the life cycle if the normal full development of some anatomical, physiological, metabolic, or psychological structure or function is to be attained. Such a period may be associated with very high susceptibility to specific potentially toxic substances.

- **critical study**
investigation yielding the no-observed adverse effect level that is used as the basis of the reference dose. See **reference dose**.

- **cross-product ratio**
see **odds ratio**.

- **cross-reactivity**
the ability of an antibody to react with or bind an antigen that did not stimulate its production.

- **cross-sectional study (of disease prevalence and associations)**
study which examines the relationship between diseases (or other health-re-

lated characteristics) and other variables of interest as they exist in a defined population at one particular time. Disease prevalence rather than incidence is normally recorded in a cross-sectional study and the temporal sequence of cause and effect cannot necessarily be determined. See **disease frequency survey, prevalence study, morbidity survey.**

- **culture**
a procedure in which material or micro-organisms from varying sources are cultivated in (culture) media for diagnostic or treatment purposes. Cultures are also performed to detect whether viable micro-organisms are present, and, therefore, an infection exists. Growth and identification of the organisms present allow suitable selection of treatment (antibiotics). The cultivation product or the material to be cultivated may also be referred to as a culture.

- **cumulative effect**
overall adverse change which occurs when repeated doses of a harmful substance or radiation have biological consequences which are mutually enhancing. See **functional accumulation**.

- **cumulative incidence ratio**
value obtained by dividing the cumulative incidence rate in the exposed population by the cumulative incidence rate in the unexposed population.

- **cumulative incidence, cumulative incidence rate**
number and proportion of a group of people who experience the onset of a health-related event during a specified time interval. This interval is generally for all members of the group, but, as in lifetime incidence, it may vary from person to person without reference to age.

- **cumulative median lethal dose**
estimate of the total administered amount of a substance which is associated with the death of half a population of animals when the substance is administered repeatedly in doses which are generally fractions of the median lethal

dose. The estimate may vary with the chosen size of the fraction (0.1, 0.2 etc.) and with the period of time over which effects are observed. It is a calculated quantity generally obtained by interpolation of available dose-response data relating the total administered amount to the response in the corresponding group of experimental animals. See **median lethal dose**

- **cutaneous**
  pertaining to the skin. See **dermal.**

- **cut-off particle diameter**
  median diameter of a range of particle sizes which will impact on a stage of a cascade impactor. It is also call 50% cut point, d_{50}, or the effective cut-off diameter.

- **cyanogenic**
  compounds able to produce cyanide. Examples are the cyanogenic glycosides such as amygdalin in peach and apricot stones.

- **cyanosis**
  bluish colouration, especially of the skin and mucous membranes and fingernail beds, caused by abnormally large amounts of reduced haemoglobin in the blood vessels as a result of deficient oxygenation.

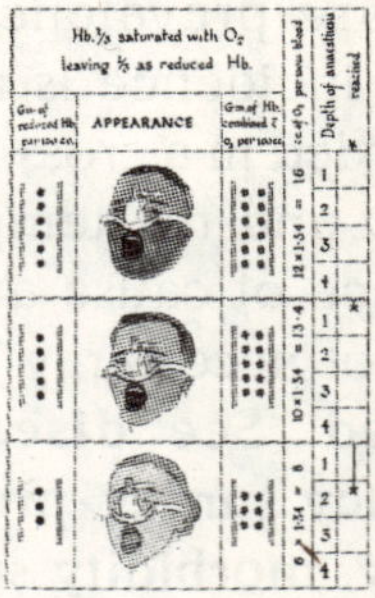

- **cyclone**
  a device in which particles are removed by centrifugal forces in a cyclonic path.

- **cytochrome**
  haemoprotein whose characteristic mode of action involves transfer of reducing equivalents associated with a reversible change in oxidation state of the haem prosthetic group. Strictly, the cytochrome P450 family are not cytochromes but haem-thiolate proteins.

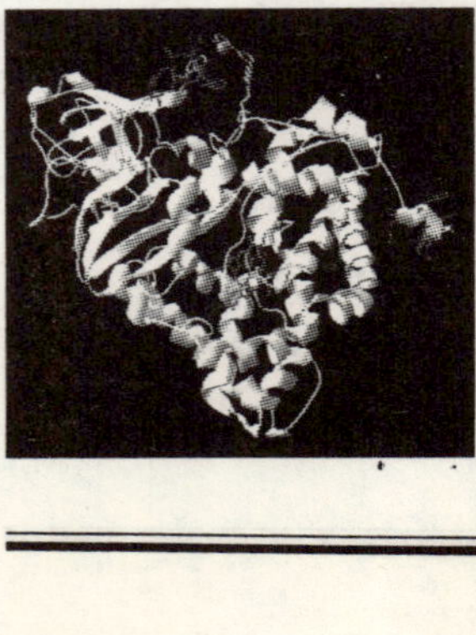

• **cytochrome p-420**
inactive derivative of cytochrome P-450 found in microsomal preparations. See **cytochrome P-448, cytochrome P-450, endoplasmic reticulum, microsome, mono-oxygenase, phase 1 reactions.**

• **cytochrome p-448**
obsolete term for cytochrome P-450 I, A1, and A2, one of the major families of the cytochromes P-450 haemoproteins. During the mono-oxygenation of certain substances, often a detoxification process, these iso-enzymes may produce intermediates which initiate mutations, chemical carcinogenesis, immunotoxic reactions and other forms of chemical toxicity. See **cytochrome P-420, cytochrome P-450, endoplasmic reticulum, microsome, mono-oxygenase, phase 1 reactions**.

• **cytochrome p-450**
haemoproteins which form the major part of the enzymes concerned with the mono-oxygenation of many endogenous and exogenous substrates. The term includes a large number of iso-enzymes which are coded for by a superfamily of genes. Endogenous substrates of these enzymes include cholesterol, steroid hormones and the eicosenoids, the exogenous substrates are xenobiotics. Strictly, the cytochrome P450 family are not cytochromes but are haemthiolate proteins.
See **mixed-function oxidase, cytochrome P-420, cytochrome P-448, endoplasmic reticulum, microsome, mono-oxygenase, phase 1 reactions, xenobiotics**.

• **cytogenetics**
branch of genetics which correlates the structure and number of chromosomes as seen in isolated cells with variation in genotype and phenotype. See **phenotype**.

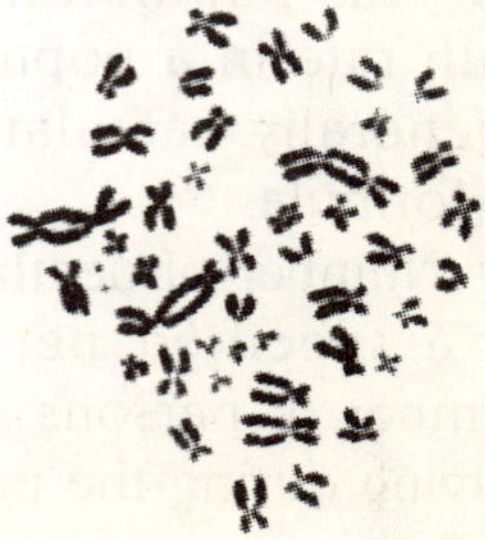

• **cytoplasm**
fundamental substance or matrix of the cell (within the plasma membrane) which surrounds the nucleus, endoplasmic reticulum, mitochondria and other organelles.

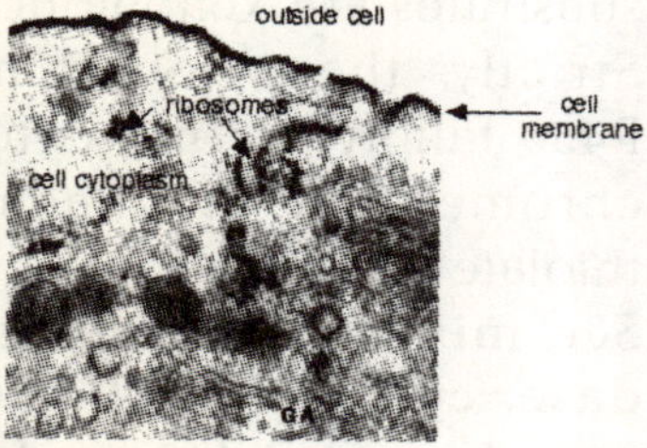

• **cytotoxic**
causing damage to cell structure or function.

• **death rate**
estimate of the proportion of a population which dies during a specified period. The numerator is the number of persons dying during the period, the denominator is the size of the population, usually estimated as the mid-year population. The death rate in a population is generally calculated by the formula:
$10n$ (Number of deaths during a specified period) / (Number of persons at risk of dying during the period)
This rate is an estimate of the person-time death rate, the death rate per $10n$ person-years: usually $n = 3$. If the rate is low, it is also a good estimate of the cumulative death rate. This rate is also called the crude death rate. See **mortality, mortality rate.**

• **de-chlorination**
removal of chlorine and chemical replacement with hydrogen or hydroxide ions to detoxify a substance.

• **decompensation**
explicit pathophysiological changes following compensation for adverse effects.

• **decontamination**
process of rendering harmless (by neutralisation, elimination, removal etc.) a potentially toxic substance in the natural environment, laboratory areas, the workplace, other indoor areas, clothes, food, water, sewage etc.

• **deep well injection**
a process by which waste fluids are injected deep below the surface of the earth.

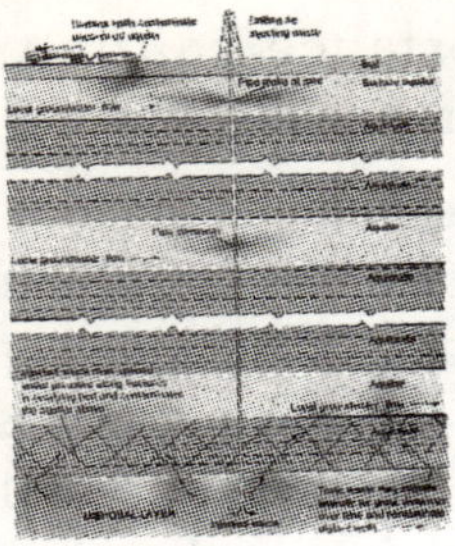

• **defoliant**
substance used for removal of leaves by its toxic action on living plants.

• **dehydrogenase**
enzyme which catalyses oxidation of compounds by removing hydrogen.

• **deionised water**
water that has had the ions removed. Used in laboratories for making reagents.

• **delayed effect**
consequence occurring after a latent period following the end of exposure to a toxic substance or other harmful environmental factor. See **latent effect**.

• **delayed health effect**
a disease or an injury that happens as a result of exposures that might have occurred in the past.

• **denaturation**
1. addition of methanol or acetone to alcohol to make it unfit for drinking.
2. change in molecular structure of proteins so that they cannot function normally, often caused by splitting of hydrogen bonds following exposure to reactive substances or heat.

• **denitrification**
reduction of nitrates to nitrites, nitrous oxides or dinitrogen (N_2) catalysed by facultative aerobic soil bacteria under anaerobic conditions.

• **density**
ratio of weight (mass) to volume of a material, usually in grams per cubic centimetre or pounds per gallon. One cc of H_2O weighs 1 g (see **specific gravity**).

• **dental fluorosis**
variety of tooth enamel malformations due to exces-

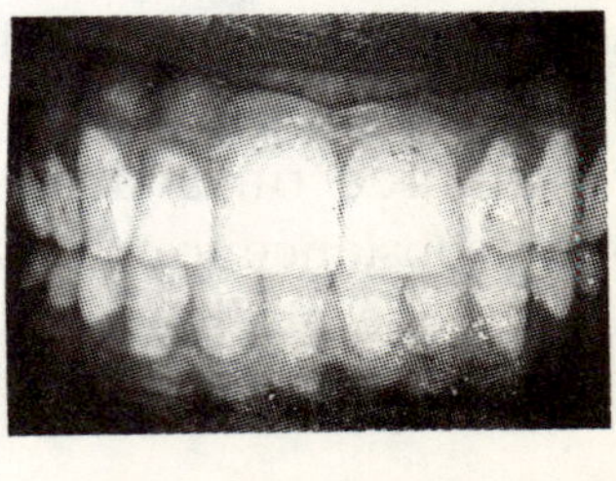

-sive fluoride exposure during dental development.

- **deoxyribonucleic acid (DNA)**
constituent of chromosomes which stores the hereditary information of an organism in the form of a sequence of purine and pyrimidine bases. This information relates to the synthesis of proteins and hence it is a determinant of all physical and functional activities of the cell, and consequently of the whole organism. See **ribonucleic acid (RNA).**

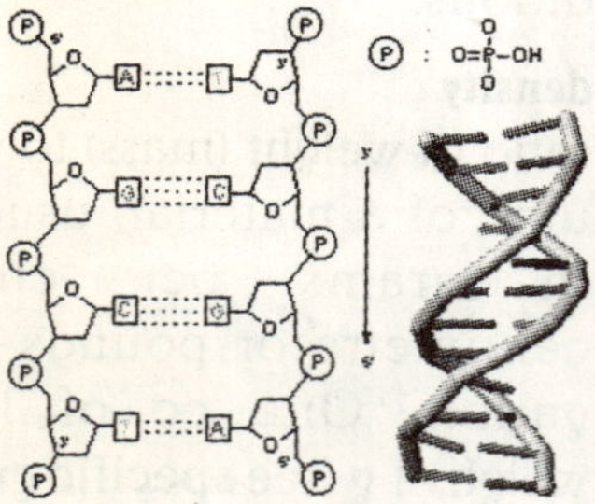

- **dependence**
1. a psychic craving for a drug or other substance which may or may not be accompanied by a physical dependency.
2. reliance on a drug or other substance to maintain health. See **addiction**.

- **depilatory**
substance causing loss of hair.

- **deposition**
1. process by which a substance arrives at a particular organ or tissue site, for example the deposition of particles on the ciliated epithelium of the bronchial airways.
2. process by which a substance sediments out of the atmosphere or water and settles in a certain place. See **accumulation**.

- **dermal**
pertaining to the skin. See **cutaneous.**

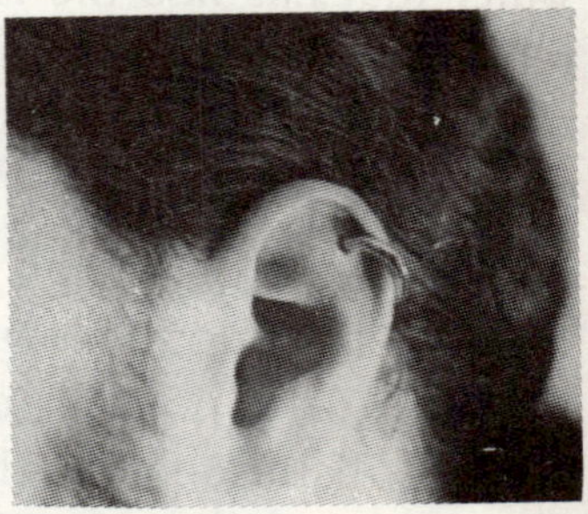

- **dermal contact**
contact with (touching) the skin (see **route of exposure)**.

- **dermal irritation**
skin reaction resulting from a single or multiple exposure to a physical or

chemical entity at the same site, characterised by the presence of inflammation. It may result in cell death.

- **dermatitis**
  inflammation of the skin. Contact dermatitis is due to local exposure and may be caused by irritation, allergy or infection.

- **descriptive epidemiology**
  study of the occurrence of disease or other health-related characteristics in populations, including general observations concerning the relationship of disease to basic characteristics such as age, sex, race, occupation, and social class. It may also be concerned with geographic location. The major characteristics in descriptive epidemiology can be classified under the headings: individuals, time and place.

- **desensitisation**
  suppression of sensitivity of an organism to an allergen to which the organism has been exposed previously.

- **desiccant**
  1. drying agent.
  2. in agriculture, a substance used for drying up plants and facilitating their mechanical harvesting.

- **desorption**
  opposite of adsorption, a decrease in the amount of adsorbed substance.

- **desquamation**
  shedding of an outer layer of skin in scales or shreds.

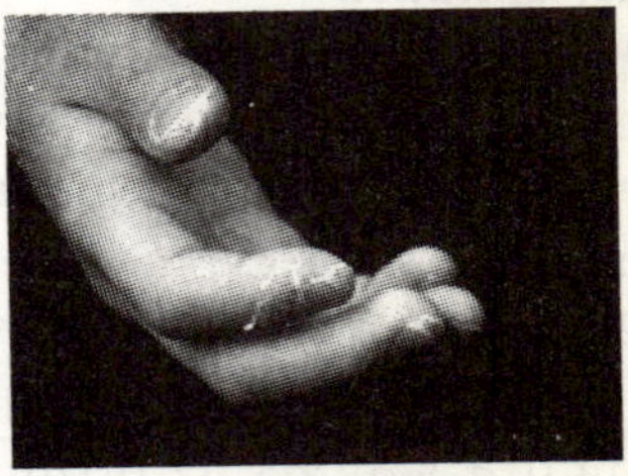

- **destruction and removal efficiency (DRE)**
  a percentage that represents the number of molecules of a compound removed or destroyed in an incinerator. A DRE of 99.99% means that 9,999 molecules are destroyed for every 10,000 that enter.

- **detection limit**
  the lowest concentration of a chemical that can reliably be distinguished from a zero concentration.

• **detergent**
a detergent is a cleaning or wetting agent, classed as anionic if it has a negative charge and cationic if it has a positive charge.

• **detoxification**
1. process, or processes, of chemical modification which make a toxic molecule less toxic.
2. treatment of patients suffering from poisoning in such a way as to promote physiological processes which reduce the probability or severity of harmful effects.

• **detoxification by haemosorption perfusion**
passage of a patient's blood through a set of columns filled with a haemosorbent (activated charcoal, ion-exchange resin, etc.). The purpose of the operation is to remove a toxic substance from the organism, particularly in an emergency. See **haemoperfusion**, **haemosorption.**

• **detoxify**
reduce the toxicity of a substance either (1) by making it less harmful or (2) by treating patients suffering from poisoning in such a way as to reduce the probability and/or severity of harmful effects.

• **detriment**
estimated measure of the expected harm or loss associated with an adverse event, usually in a manner chosen to facilitate meaningful addition over different events. It is generally the integrated product of arbitrary values of risk and hazard and is often expressed in terms such as costs in rupees, loss in expected years of life or loss in productivity, and is needed for numerical exercises such as cost-benefit analysis.

• **developmental toxicity**
adverse effects occuring at the embryo or developing foetus stage of human development.

- **dexamethasone suppression test (DTST)**
  a procedure to evaluate the neuro-regulatory mechanism of cortisol secretion into blood plasma following the administration of dexamethasone (a synthetic steroid which normally causes decreased secretion). Cortisol levels may be determined by various immunoassay methods. DTST is often done to help identify a depression disease, as patients may not suppress cortisol secretion after dexamethasone administration.

- **diabetes mellitus**
  a chronic syndrome of impaired carbohydrate, protein, and fat metabolism owing to insufficient secretion of insulin or target tissue insulin resistance, characterised by polydipsia, polyuria and polyphagia.

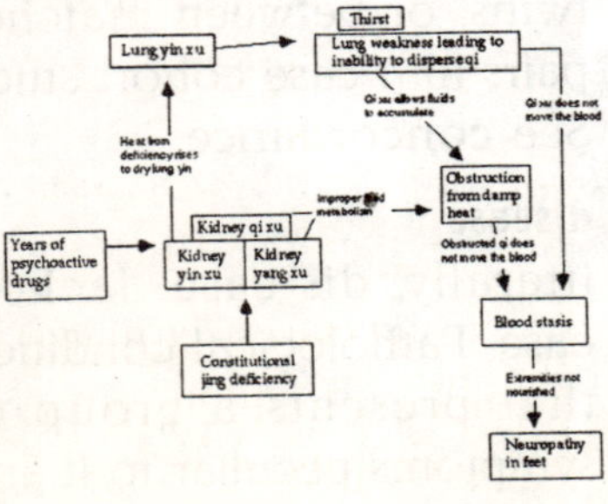

- **diaphoresis**
  profuse perspiration.

- **diaphoretic**
  causing profuse perspiration. See **sudorific**.

- **digital rectal examination**
  a common screening procedure for prostate cancer, whereby a physician inserts a gloved, lubricated finger into the rectum, in order to feel the size and shape of the prostate through the rectal wall.

- **digoxin**
  a drug commonly used to treat congestive heart failure and certain heartbeat irregularities that is obtained from leaves of the Foxglove (digitalis) plant. Digoxin overdose can be toxic. The level of the drug can be obtained from blood testing, which can aid in setting appropriate dose amounts.

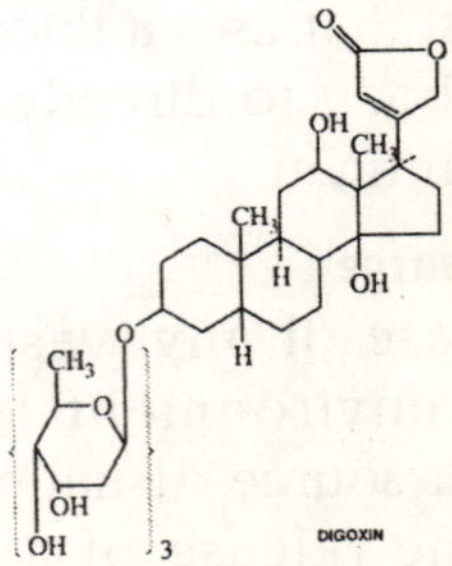

• **dilution ratio**

1. the relationship between the volume of water in a stream and the volume of incoming waste. It can affect the ability of the stream to assimilate waste.

2. to mix sample with water, saline, or buffer to decrease the concentration.

• **dimercaprol**

see **2,3-dimercaptopropan-1-ol**.

• **diploid**

chromosome state in which the chromosomes are present in homologous pairs. Normal human somatic (non-reproductive) cells are diploid (they have 46 chromosomes), whereas reproductive cells, with 23 chromosomes, are haploid. See **haploid, meiosis, mitosis**.

• **direct fluorescent antibody test (DFA)**

a test that uses a fluorescent antibody to directly detect an antigen.

• **discharge**

release of any waste into the environment from a point source. Usually refers to the release of a liquid waste into a body of water through an outlet such as a pipe, but also refers to air emissions.

• **discharge (effluent, emission) standard or release limit**

maximum amount of a pollutant released from a given source to a specified medium which is acceptable under specified circumstances.

• **discharge area**

an area of land where there is a net annual transfer of water from the ground water to surface water, such as to streams, springs, lakes, and wetlands.

• **discontinuous effect**

see **intermittent effect**.

• **discordance (genetic)**

any difference in a character between individuals due to genetic differences such as may occur in dizygotic twins, or between matched pairs in a case cohort study. See **concordance**.

• **disease**

literally, dis-ease, lack of ease. Pathological condition that presents a group of symptoms peculiar to it and which establishes the condition as an abnormal entity different from other normal or pathological body states.

- **disease prevention**
  measures used to prevent a disease or reduce its severity.
- **disease registry**
  a system of ongoing registration of all cases of a particular disease or health condition in a defined population.
- **disinfection**
  application of energy or chemical to kill or inactivate pathogenic organisms.

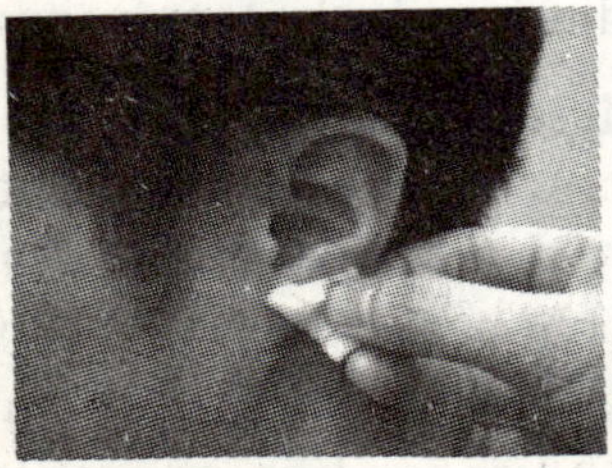

- **dispersion model**
  a mathematical prediction of how pollutants from a discharge or emission source will be distributed in the surrounding environment under given conditions of wind, temperature, humidity, and other environmental factors.
- **disposal**
  the discharge, deposit, injection, dumping, spilling, leaking, or placing of any solid waste or hazardous waste into the environment (land, surface water, ground water, and air).

- **disposal facility**
  a landfill, incinerator, or other facility which receives waste for disposal. The facility may have one or many disposal methods available for use. Does not include wastewater treatment.
- **disposition**
  natural tendency shown by an individual or group of individuals, including any tendency to acquisition of specific diseases, often due to hereditary factors.

- **dissipation**
reduction in the amount of a pesticide or other compound which has been applied to plants, soil etc. (used when it is not clear whether this is by mineralisation degradation, binding, or leaching).

- **dissolved oxygen (DO)**
oxygen that is freely available in water to sustain the lives of fish and other aquatic organisms.

- **distributed source**
see **area source, point source**.

- **distribution**
1. dispersal of a substance and its derivatives throughout the natural environment.
2. dispersal of a substance within an organism, including metabolism, storage and excretion.

- **diuresis**
excretion of urine, especially in excess.

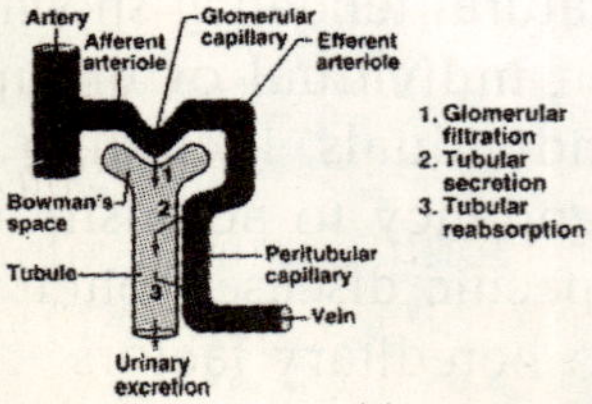

- **diuretic**
agent which increases urine production. See **micturitic**.

- **dosage**
dose expressed as a function of the organism being dosed and time, for example mg/(kg body weight)/day. See **dose**.

- **dose**
total amount of a substance administered to, taken or absorbed by an organism. See **absolute lethal dose, cumulative median lethal dose, lethal dose, maximum tolerable dose, median effective dose**,.

- **dose (for radioactive chemicals)**
the radiation dose is the amount of energy from radiation that is actually absorbed by the body. This is not the same as measurements of the amount of radiation in the environment.

- **dose response**
how an organism's response to a toxic substance changes as its overall exposure to the substance changes. For example, a small dose of carbon monoxide may cause drowsi-

ness, a large dose can be fatal.

- **dose-effect curve**
graph of the relation between dose and the magnitude of the biological change produced measured in appropriate units. See **concentration-effect curve**.

- **dose-effect relationship**
association between dose and the magnitude of a continuously graded effect, either in an individual or in a population or in experimental animals.
See **concentration-effect relationship**.

- **dose-related effect**
situation in which the magnitude of a biological change is related to the dose. See **non-dose-related effect**.

- **dose-response curve**
graph of the relation between dose and the proportion of individuals in a population responding with an all-or-none effect. See **concentration-response curve, response**.

- **dose-response relationship**
association between dose and the incidence of a defined biological effect in an exposed population. See **concentration-response relationship, response.**

- **draize test**
evaluation of materials for their potential to cause dermal or ocular irritation and corrosion following local exposure. It is generally done using the rabbit model (almost exclusively the New Zealand White) although other animal species have been used.

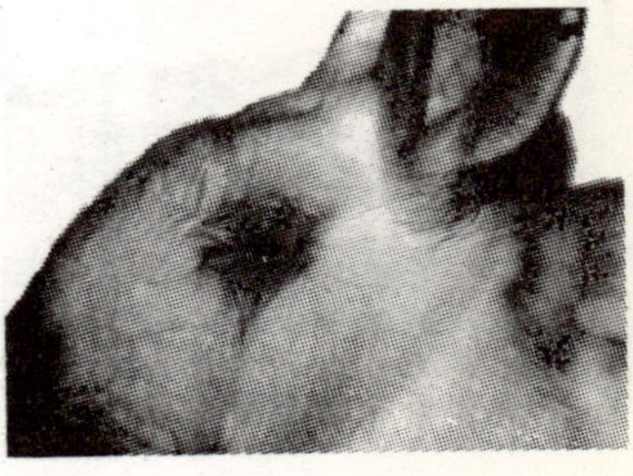

- **drug**
any substance which when absorbed into a living organism may modify one or more of its functions. The term is generally accepted for a substance taken for a therapeutic purpose, but is also commonly used for abused substances. See **medicine, pharmaceutical.**

- **dump**
a land site where wastes are discarded in a disorderly or haphazard fashion without regard to protecting the environment. Uncontrolled dumping is an indiscriminate and illegal form of waste disposal. Problems associated with dumps include multiplication of disease-carrying organisms and pests, fires, air and water pollution, unsightliness, loss of habitat, and personal injury.

- **duplicate (replicate) samples (in chemistry)**
two (or multiple) samples taken under the same or comparable conditions.

- **duplicate portion sampling method (diet/food)**
method frequently used for the same purposes as the total diet study technique. Test persons consume their ordinary diet but for each meal, they prepare for subsequent analysis a duplicate portion of all food as prepared, served and consumed. See **duplicate diet study**.

- **dust**
an aerosol consisting of solid particles made airborne by the mechanical disintegration of bulk solid material (e.g., during cutting, crushing, grinding, abrasion, transportation, etc.), with sizes ranging from as low as sub-micrometer (μm) to over 100μm.

- **dysarthria**
imperfect articulation of speech due to neuromuscular damage.

- **dysfunction**
abnormal, impaired, or incomplete functioning of an organism, organ, tissue or cell.

- **dysplasia**
  abnormal development of an organ or tissue identified by morphological examination.

- **dyspnoea**
  difficult or laboured breathing.

- **echocardiography**
  a non-invasive procedure that utilises reflected ultrasonic waves (echoes) directed to the heart that are converted and recorded as an image showing internal structures. It is used to study the heart structure and motion in diagnosis of pathologic cardiovascular changes.

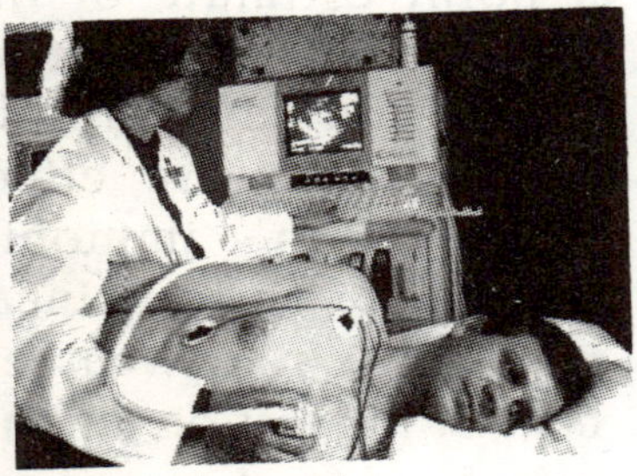

- **ECn**
  this is the commonly used abbreviation for the exposure concentration of a toxicant causing a defined effect on n% of a test population.

- **ecogenetics**
  study of the influence of hereditary factors on the effects of xenobiotics on individual organisms. See **pharmacogenetics**, **toxicogenetics**, **polymorphism**.

- **ecology**
  branch of biology which studies the interactions between living organisms and all factors (including other organisms) in their environment. Such interactions encompass environmental factors which determine the distributions of living organisms.

- **ecosystem**
  1. grouping of organisms (micro-organisms, plants, animals) interacting together, with and through their physical and chemical environments, to form a functional entity.
  2. the interacting synergism of all living organisms in a particular environment, every plant, insect, aquatic animal, bird, or land species that forms a complex web of interdependency. An action taken at any level in the food

chain, use of a pesticide for example, has a potential domino effect on every other occupant of that system.

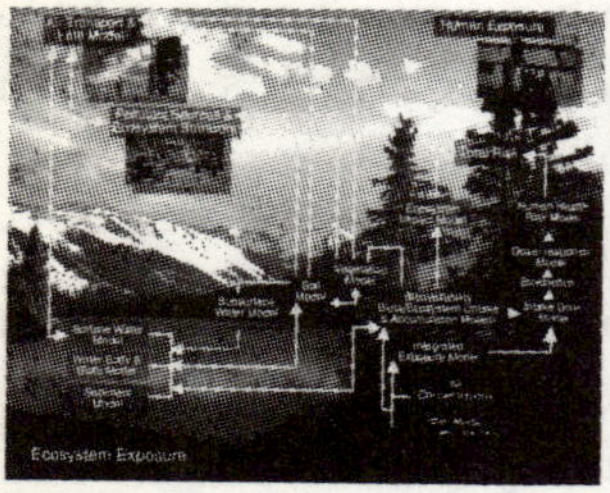

- **ecotoxicology**
  study of the toxic effects of chemical and physical agents on all living organisms, especially on populations and communities within defined ecosystems. It includes transfer pathways of these agents and their interactions with the environment.
- **ectohormone**
  see **pheromone**.
- **ectoparasiticide**
  substance intended to kill parasites living on the exterior of the host.
- **eczema**
  acute or chronic skin inflammation with erythema, papules, vesicles, pustules, scales, crusts or scabs, alone or in combination, of varied aetiology.

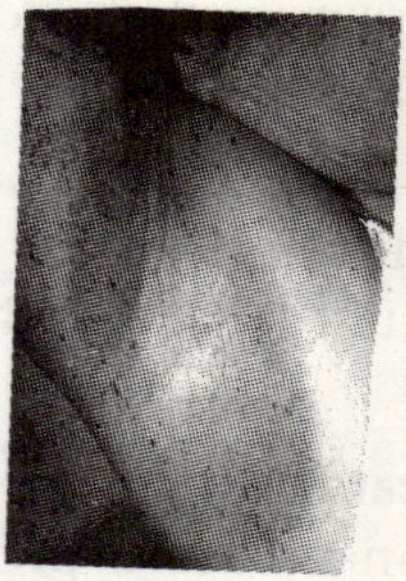

- **edema**
  see **oedema**.
- **EDn**
  this is the commonly used abbreviation for the dose of a toxicant causing a defined effect on n% of a test population.
- **effect concentration (EC)**
  a point estimate of the toxicant concentration that would cause an observable adverse effect (e.g., death, immobilisation, or serious incapacitation) in a given percent of the test organisms, calculated from a continuous model (e.g., Probit Model). EC25 is a point estimate of the toxicant concentration that would cause an observable adverse effect in 25 percent

of the organisms (EPA, 2000). The lower the EC25, the more toxic the chemical or effluent sample. Other EC values, e.g. the EC50 or EC10 may also be calculated to determine concentrations causing effect in a greater or lesser proportion of the population.

- **effective concentration (EC)**
concentration of a substance that causes a defined magnitude of response in a given system: EC_{50} is the median concentration that causes 50 % of maximal response. See **lethal concentration.**

- **effective dose (ED)**
dose of a substance that causes a defined magnitude of response in a given system. ED_{50} is the median dose that causes 50 % of maximal response. See **dose, lethal dose**.

- **effluent**
fluid, solid or gas discharged from a given source into the external environment. See **emission.**

- **effluent limitations**
limits on the amounts of pollutants which may be discharged by a facility. These limits are calculated so that water quality standards will not be violated even at low stream flows.

- **ejaculation**
a sudden release of fluid, especially of semen, from the body.

- **electrical conductivity**
property which allows an electric current to flow when a potential difference is applied. It is the reciprocal of the resistance in ohms measured between opposite faces of a centimetre cube of an aqueous solution at a specified temperature. It is expressed as micromho per centimetre

- **electroblot tests**
a number of diagnostic transfer methods (blots) which allow shifting of proteins from gels to cellulose, where they are isolated and immobilised and can be detected and identified by a variety of techniques. The

cellulose soaks up proteins like a blotter.

- **electrocardiogram (ECG)**
a graphic recording that shows electrical activity generated by the heart muscle. It is used to help diagnose specific cardiac anomalies such as disturbances of heart rhythm and conduction.

- **electrocardiograph**
the instrument used to detect and record the electrical activity of the heart muscle. It produces the ECG.

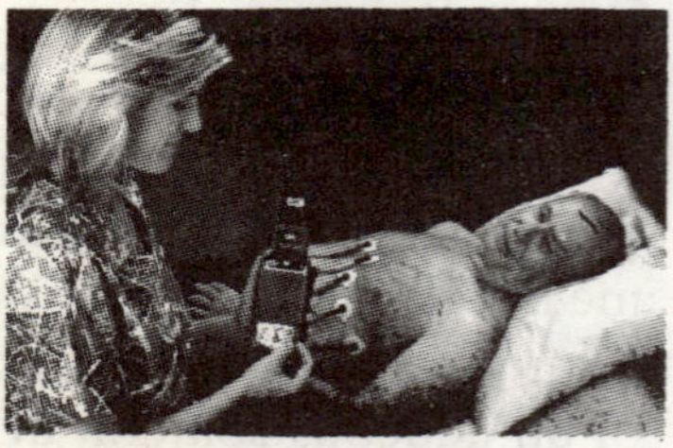

- **electrodes**
conductive materials, placed in water solution, which have a positive or negative charge.

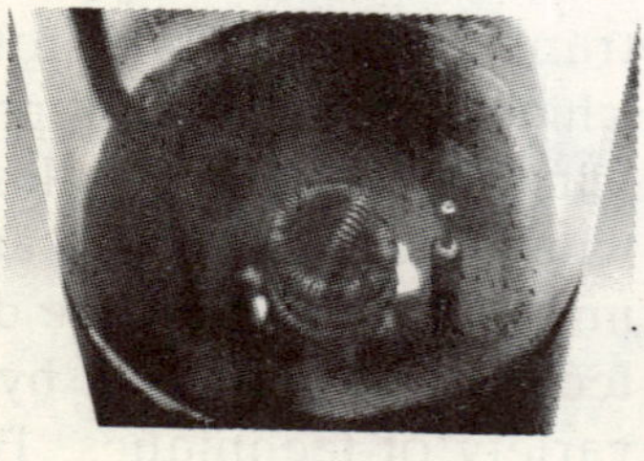

- **electrodialysis**
a process that uses electrical current applied to permeable membranes to remove minerals from water. Often used to desalinise salt or brackish water.

- **electroencephalogram (EEG)**
a chart that provides a record of the electrical activity of the brain. which is derived from electrodes placed on the scalp. The instrument used is an electroencephalograph. An electroencephalogram is useful in diagnosis of seizure disorders and tumour detection.

- **electrolyte**
1. a solution of a compound that dissociates into ions and can conduct electricity. Electrolytes affect the movement of substances between body fluids and tissues, and are crucial for normal function and metabolism. In many illnesses, monitoring proper balance among the most important ones is vital. Calcium and potassium, for example, are necessary for cardiac and skeletal contractions and relaxations, and sodium is essential for

fluid balance. An electrolyte imbalance can be caused by disease conditions or medications. Electrolytes are also used as electrical conductors in electrophoretic test methods.

2. a non-metallic substance that conducts an electric current in solution by the movement of ions rather than electrons.

- **electron microscope**

an electron-optical instrument that utilises a beam of electrons, rather than light,

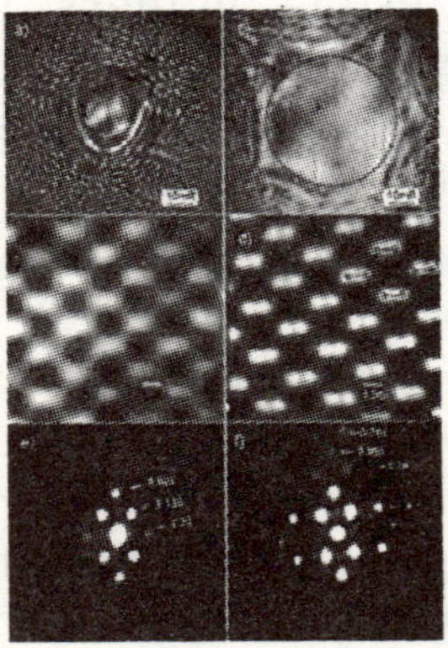

to focus on cell surfaces of a very thin specimen to produce an enlarged image on a fluorescent screen or photographic plate. Because resolution (the ability to distinguish adjacent objects as separate) is better and magnification 1,000 times that of an optical light microscope is possible, electron microscopy can help determine the nature of tumours and kidney disease. See **scanning electron microscope**.

- **electroneutrality**

where the positive charges equal the negative charges.

- **electronic foetal monitoring**

the use of external and internal devices that record the foetal heart rate. An externally applied ultrasound transducer converts sound energy into electrical energy which is recorded on a graph. After rupture of the membranes, an electrode can be applied via the maternal vagina to the foetal scalp to obtain a foetal ECG. The maternal uterine contractions also are recorded on the graph so that they can be related to the foetal heart rate, to indicate foetal distress.

- **electrophoresis**

movement of charged particles in an electrical field, a technique used to separate materials in a substance by placing a sample in a buffer solution through which an electrical current is then passed. Extent of migration and separation of components depends on molecular weight and electrical charge of the particles. This method is often used to determine the proportions of various proteins in serum.

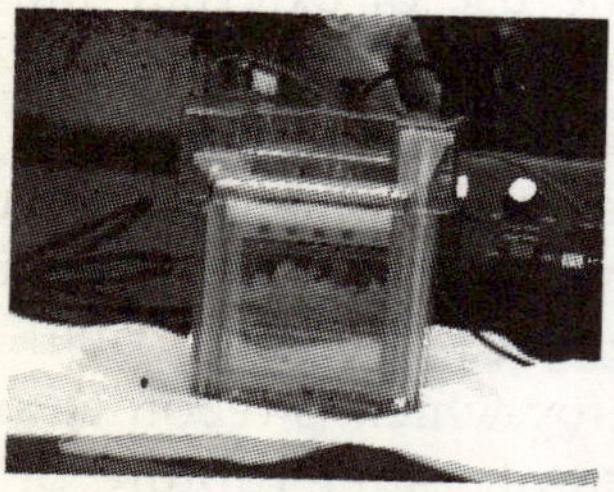

- **electrostatic precipitator (ESP)**
  an air pollution control device that removes particles from a gas stream (smoke) after combustion occurs. The ESP imparts an electrical charge to the particles, causing them to adhere to metal plates inside the precipitator. Rapping on the plates causes the particles to fall into a hopper for disposal.

- **element**
  a substance which cannot be separated into its constituent parts and still retain its chemical identity. For example, sodium (Na) is an element.

- **elimination**
  the process whereby a substance or other material is expelled from the body (or a defined part thereof), usually by a process of extrusion or exclusion but sometimes through metabolic transformation. The combination of chemical degradation of a xenobiotic in the body and excretion by the intestine, kidneys, lungs, skin, in sweat, expired air, milk, semen, menstrual fluid or secreted fluids.

- **elimination half-life or half time**
  period taken for the plasma concentration of a substance to decrease by half. See **biological half-life or half-time($T_{1/2}$)**.

- **eliminator (of a poison)**
  substance that contributes to the elimination of a poison from an organism.

- **elutriate**
  an aqueous sediment extract prepared by decanting the overlying water

from a sediment-water mixture, presumably representing the labile, bioavailable components of the sediment.

• **embryo**
1. stage in the developing mammal at which the characteristic organs and organ systems are being formed. For humans, this involves the stages of development from the second to the eighth week (inclusive post conception).

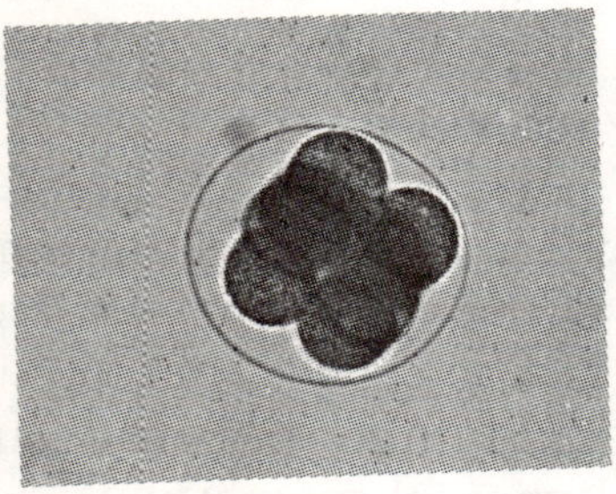

2. in birds, the stage of development from the fertilization of the ovum up to hatching.
3. in plants, the stage of development within the seed.

• **embryotoxic**
this adjective is applied to any substance which is harmful, in any sense, to an embryo.

• **embryotoxicity**
production by a substance of toxic effects in progeny in the first period of pregnancy between conception and the foetal stage. See **developmental toxicity, teratogenicity.**

• **embryotropic effect**
change in the embryo and the regulation of its development.

• **emesis**
vomitting.

• **emission**
release of a substance from a source, including discharges to the wider environment. See **discharge, effluent, release, immission.**

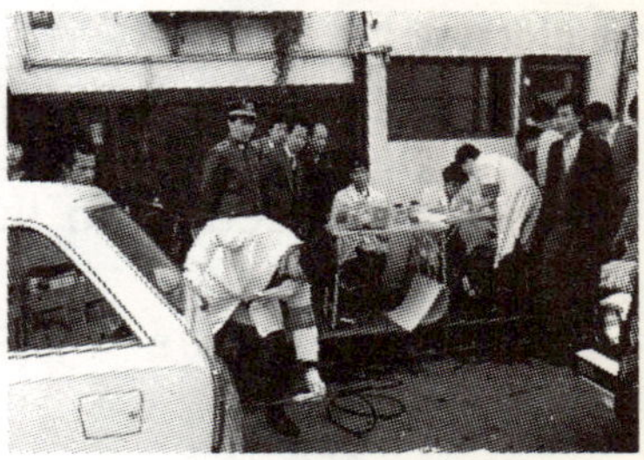

• **emission and exposure control**
technical and administrative procedures and specifications applied for the monitoring, reduction or elimination of emissions from a source or exposure to a target.

- **emission standard**
  quantitative limit on the emission or discharge of a substance from a source, usually expressed in terms of a time-weighted average concentration or a ceiling value. See **discharge standard**, **limit value.**

- **endangered species**
  animals, plants, birds, fish, or other living organisms threatened with extinction by man-made or natural changes in the environment.

- **endemic**
  present in a community or among a group of people, said of a disease prevailing continually in a region.

- **endocrine**
  pertaining to hormones or to the glands that secrete hormones directly into the bloodstream.

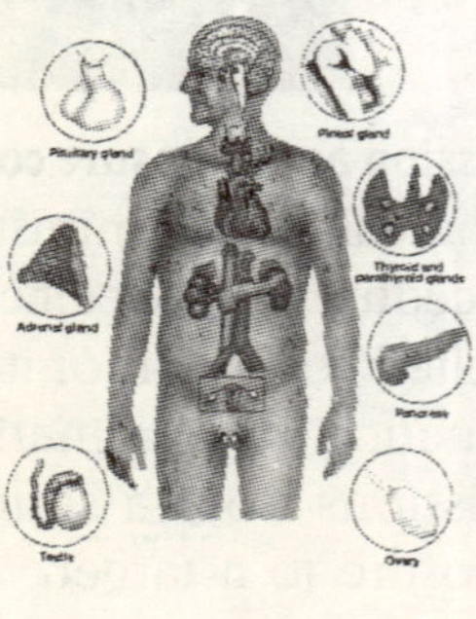

- **endoplasmic reticulum**
  intracellular complex of membranes in which proteins and lipids, as well as molecules for export, are synthesised and in which the bio-transformation reactions of the mono-oxygenase enzyme systems occur. It may be isolated as microsomes following cell fractionation procedures. See **cytochrome P-420, cytochrome P-448, cytochrome P-450, microsome, mono-oxygenase, phase 1 reactions**.

- **endothelial**
  pertaining to the layer of flat cells lining the inner surface of blood and lymphatic vessels, and the surface lining of serous and synovial membranes.

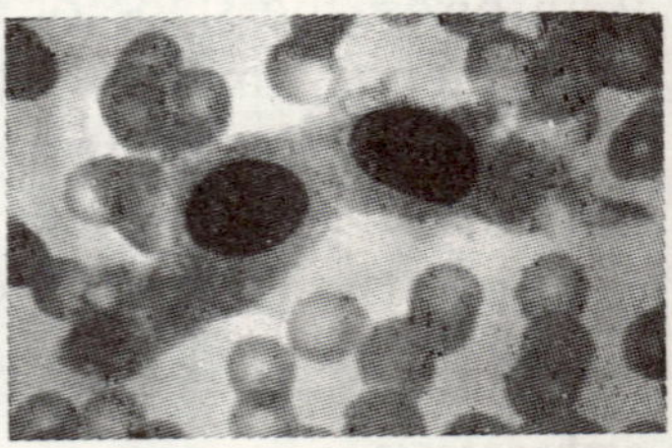

- **energy recovery**
  to capture energy from waste through any of a variety of processes (e.g., burning). Many new technology

incinerators are waste-to-energy recovery units.

- **enteritis**
intestinal inflammation.

- **enterohepatic circulation**
cyclical process involving intestinal re-absorption of a substance that has been excreted through the bile followed by transfer back to the liver, making it available for biliary excretion again.

- **environment**
aggregate, at a given moment, of all external conditions and influences to which a system under study is subjected.

- **environmental audit**
this is an account by manufacturers of the products produced and their effects on the environment- energy use policy, materials use policy, waste output and their effect on the environment. Purchasing procedure monitoring to obtain environmentally sound goods and services, etc. may also be adopted as part of an environmental audit.

- **environmental damage**
adverse effects to the natural environment.

- **environmental equity**
equal protection from environmental hazards for individuals, groups, or communities regardless of race, ethnicity, or economic status.

- **environmental exposure level (EEL)**
level (concentration or amount or a time integral of either) of a substance to which an organism or other component of the environment is exposed in its natural surroundings.

- **environmental fate**
destiny of a chemical or biological pollutant after release into the natural environment.

- **environmental health**
human welfare and its change by the environment, including technical and administrative measures for improving the human environment from a health point of view. See **environmental medicine, environmental hygiene, occupational hygiene**.

- **environmental health impact assessment**
estimate of the adverse health effects or risks likely to follow from a proposed or expected environmental change or development.

- **environmental hygiene**
practical control measures used to improve the basic environmental conditions affecting human health, for example, clean water supply, human and animal waste disposal, protection of food from biological contamination, and housing conditions, all of which are concerned with the quality of the human environment. See **environmental sanitation**.

- **environmental impact assessment (EIA)**
appraisal of the possible environmental consequences of a past, present, or planned action, resulting in the production of an environmental impact statement or 'finding of no significant impact (FONSI)'. See **environmental impact statement**.

- **environmental impact statement (EIS)**
report resulting from an environmental impact assessment. See **environmental impact assessment**.

- **environmental justice**
the fair treatment of people of all races, cultures, incomes, and educational levels with respect to the development and enforcement of environmental laws, regulations, and policies. Fair treatment implies that no population should be forced to shoulder a disproportionate share of exposure to the negative effects of pollution due to lack of political or economic strength.

- **environmental media**
soil, water, air, biota (plants and animals), or any other parts of the environment that can contain contaminants.

- **environmental media and transport mechanism**
environmental media include water, air, soil, and biota (plants and animals). Transport mechanisms move contaminants from the source to points where

human exposure can occur. The environmental media and transport mechanism is the second part of an exposure pathway.

- **environmental monitoring**
continuous or repeated measurement of agents in the environment to evaluate environmental exposure and possible damage by comparison with appropriate reference values based on knowledge of the probable relationship between ambient exposure and resultant adverse effects. See **biological effect monitoring, biological monitoring, reference value.**

- **environmental protection**
actions taken to prevent or minimise adverse effects to the natural environment.
Complex of measures including monitoring of environmental pollution, development and practice of environmental protection principles (legal, technical, and hygienic), including risk assessment, risk management and risk communication.

- **environmental quality objective (EQO)**
overall state to be aimed for in a particular aspect of the natural environment, for example, 'water in an estuary such that shellfish populations survive in good health'. Unlike an environmental quality standard, the EQO is usually expressed in qualitative and not quantitative terms. See **environmental quality standard.**

- **environmental quality standard (EQS)**
amount concentration or mass concentration of a substance that should not be exceeded in an environmental system, often expressed as a time-weighted average measurement over a defined period. See **limit value**.

- **environmental sanitation**
see **environmental hygiene**.

- **environmental toxicology**
the study of the toxicology of those substances or physical agents which are encountered in the environment.

- **environmental transformation**
chemical transformation of substances resulting from interactions in the environment.

- **enzootic**
present in a community or among a group of animals, said of a disease prevailing continually in a region.

- **enzyme**
a protein produced by cells that influences the speed of a specific chemical reaction without being used up or permanently changed in the process. Enzymes are catalysts.

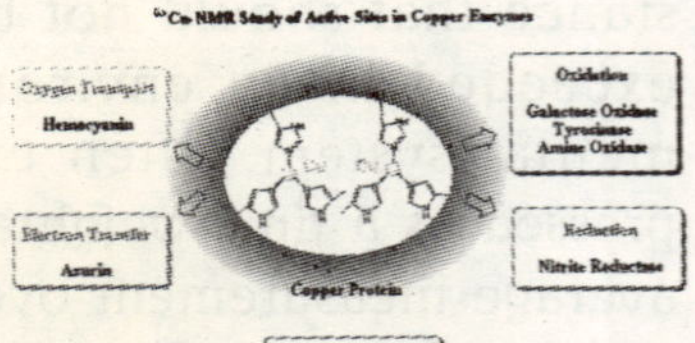

- **enzyme immunoassay (EIA)**
an assay that uses an enzyme-bound antibody to detect antigen. The enzyme catalyses a colour reaction when exposed to substrate.

- **enzyme-linked fluorescent assay (ELFA)**
similar to EIA except that the enzyme catalyses a fluorescence, not a colour reaction.

- **enzyme-linked immunosorbent assay (ELISA)**
see assay.

- **enzymic (enzymatic) process**
any chemical reaction or series of reactions catalysed by an enzyme or enzymes is described as an enzymic or enzymatic process. An enzyme is a protein which acts as a highly selective catalyst permitting reactions to take place rapidly in living cells under physiological conditions.

- **epidemiologist**
a medical scientist who studies the various factors involved in the incidence, distribution, and control of disease in a population.

- **epidemiology**
study of the distribution and determinants of health-related states or events in populations and the application of this study to control of health problems.

- **epigastric**
pertaining to the upper-middle region of the abdomen.

• **epigenesis**
changes in an organism brought about by alterations in the expression of genetic information without any change in the genome itself: the genotype is unaffected by such a change but the phenotype is altered. See **mutation**, **phenotype**, **transformation**, **tumour**.

• **epigenetic changes**
any changes in an organism brought about by alterations in the action of genes are called epigenetic changes. Epigenetic transformation refers to those processes which cause normal cells to become tumour cells without any mutations having occurred. See **mutation**, **transformation**, **tumour**.

• **epileptiform**
occurring in severe or sudden spasms, as in convulsion or epilepsy.

• **epithelioma**
any tumour derived from epithelium. See **carcinoma**.

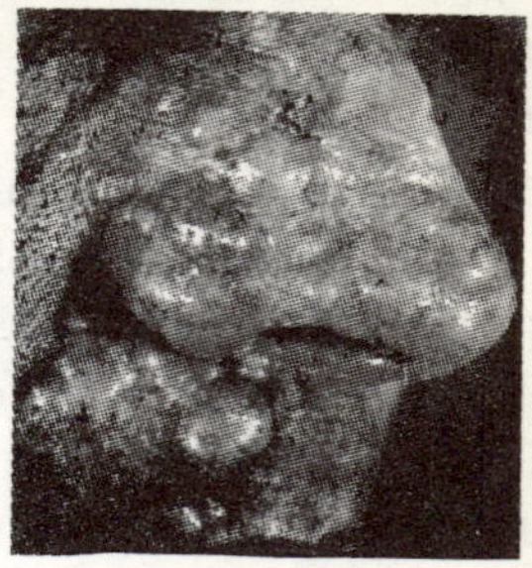

• **epithelium**
cells covering the internal and external surfaces of the body.

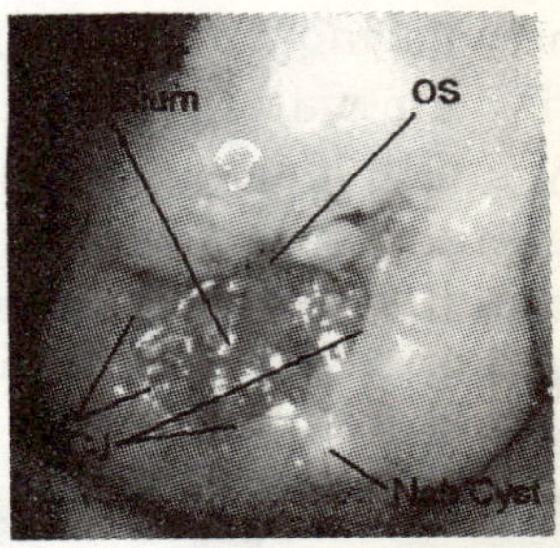

• **epitope**
any part of a molecule that acts as an antigenic determinant: a macromolecule can contain many different

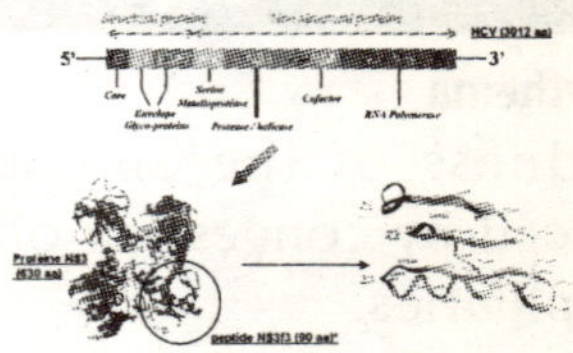

epitopes each capable of stimulating production of a different specific antibody.

- **equivalent diameter (of a particle)**
diameter of a spherical particle of the same density as a particle under investigation that, relative to a given phenomenon or property, would behave in the same way as the particle under investigation. See **aerodynamic diameter**.

- **erosion**
the wearing away of soil by wind or water, intensified by land-clearing practices related to farming, residential or industrial development, road building, or logging.

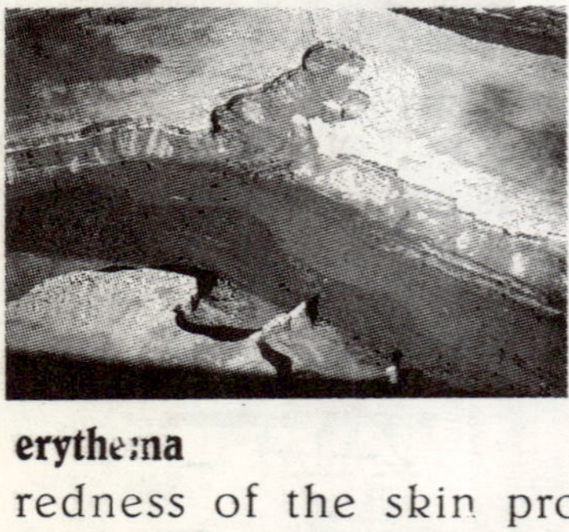

- **erythema**
redness of the skin produced by congestion of the capillaries.

Erythema migrans lesions

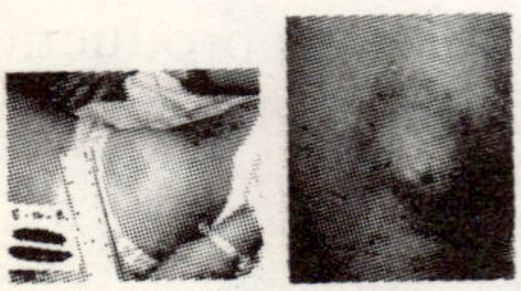

- **eschar**
slough or dry scab on an area of skin that has been burnt.

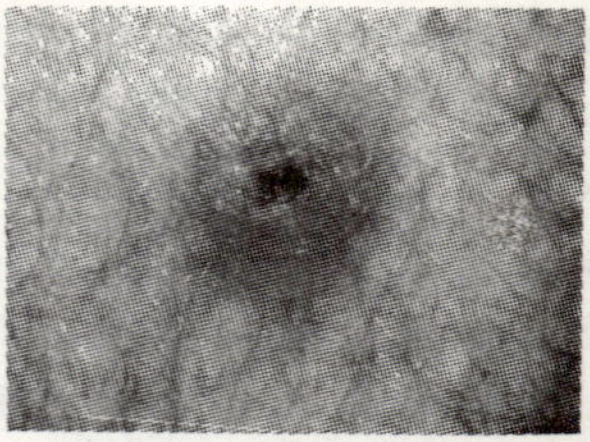

- **estimated daily intake (EDI)**
prediction of the daily intake of a residue of a potentially harmful agent based on the most realistic estimation of the residue levels in food and the best available food consumption data for a specific population. Residue levels are estimated taking into account known uses of the agent, the range of contaminated commodities, the proportion of a commodity treated, and the quantity of home-grown or imported commodities. The EDI is expressed in mg residue per person.

- **estimated exposure concentration (EEC)**
measured or calculated amount or mass concentration of a substance to which an organism is likely to be

exposed, considering exposure by all sources and routes.

- **estimated exposure dose (EED)**
  measured or calculated dose of a substance to which an organism is likely to be exposed, considering exposure by all sources and routes.

- **estimated maximum daily intake (EMDI)**
  prediction of the maximum daily intake of a residue of a potentially harmful agent based on assumptions of average food consumption per person and maximum residues in the edible portion of a commodity, corrected for the reduction or increase in residues resulting from preparation, cooking, or commercial processing. The EMDI is expressed in mg residue per person.

- **estrogen**
  a female sex hormone.

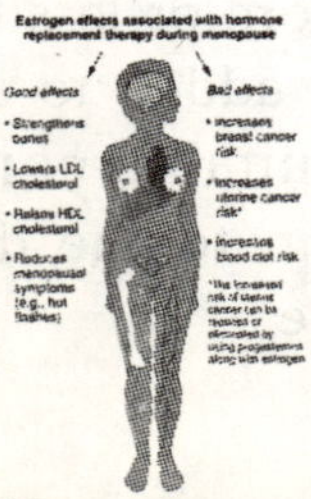

- **estuary**
  a complex ecosystem between a river and near-shore ocean waters where fresh and salt water mix. These brackish areas include bays, mouths of rivers, salt marshes, wetlands, and lagoons and are influenced by tides and currents. Estuaries provide valuable habitat for marine animals, birds, and other wildlife.

- **ET50**
  this abbreviation is used for the exposure time required to produce a defined effect when a test population is exposed to a fixed concentration or specified dose of a toxicant.

- **ethanol**
  ethyl alcohol, the type used in alcoholic beverages, medicines, and food products. Blood levels of ethanol can be ascertained by tests

that indicate the degree of intoxication. Alcohol affects brain function (central nervous system depressant) and may produce chronic illness in susceptible individuals.

- **etiology**
see **aetiology**.

- **eukaryote**
cell or organism with the genetic material packed in a membrane-surrounded structurally discrete nucleus and with well-developed cell organelles. The term includes all organisms except archaebacteria, eubacteria and cyanobacteria (until recently classified as cyanophyta or blue-green algae). See **prokaryote**.

- **euthyroidism**
a normal thyroid hormone status most precisely defined by a normal TSH level.

- **eutrophic**
describes a body of water with a high concentration of nutrient salts and a high or excessive rate of biological production.

- **eutrophication**
eutrophication refers to the accumulation of nutrients in a lake or land-locked body of water. This occurs naturally over many years but has recently been accelerated by fertiliser runoff from farms and sewage input. Algal blooms result and their decay removes dissolved oxygen, eliminating aerobic organisms such as fish, and may cause accumulation of sulphide in the water.

- **excess lifetime risk**
additional or excess risk incurred over the lifetime of an individual by exposure to a toxic substance. See **hazard**.

- **excess rate**
see **rate difference**.

- **exchange transfusion**
method of active artificial elimination of toxicity consisting in complete replacement of blood of the patient by donor blood.

- **excipient**
any more or less inert substance added to a drug to give suitable consistency or form to the drug. See **vehicle**.

• **excretion**
discharge or elimination of an absorbed or endogenous substance, or of a waste product, and/or their metabolites, through some tissue of the body and its appearance in urine, faeces, or other products normally leaving the body. Excretion of chemical compounds from the body occurs mainly through the kidney and the gut. Volatile compounds may be largely eliminated by exhalation. Excretion by perspiration and through hair and nails may also occur. Excretion by the gastrointestinal tract may take place by various routes such as the bile, the shedding of intestinal cells and transport through the intestinal mucosa. See **clearance, elimination.**

• **excretion rate**
amount of substance (and/or its metabolites) or fraction that is excreted per unit time. It should be noted that according to this definition excretion does not include the passing of a substance through the intestine without absorption. When discussing the total amount of a substance in faeces (including the unabsorbed part), it is preferable to speak about faecal substance content (mol/kg) or mass content (kg/kg).

• **exogenous**
resulting from causes or derived from materials external to an organism. See **endogenous**.

• **exogenous substance**
see **xenobiotic.**

• **experimental model ecosystem**
see **microcosm.**

• **explant**
living tissue removed from its normal environment and transferred to an artificial medium for growth.

• **explosive limits (chemical)**
the amounts of vapour in air that form explosive mixtures. These limits are expressed as lower and upper values and give the range of vapour concentrations in air that will explode if an ignition source is present.

• **exposed**
subject to a factor that is under study in the environment, for instance an envi-

ronmental hazard. See **non-exposed**, **unexposed**.

- **exposed group (sometimes abbreviated to exposed) in epidemiology**
  group whose members have been exposed to a supposed cause of a disease or health state of interest, or possess a characteristic that is a determinant of the health outcome of interest.

- **exposure**
  process by which a substance becomes available for absorption by the target population, organism, organ, tissue or cell, by any route.

- **exposure assessment**
  process of measuring or estimating concentration (or intensity), duration and frequency of exposures to an agent present in the environment or, if estimating hypothetical exposures, that might arise from the release of a substance, or radionuclide, into the environment. See **risk assessment**.

- **exposure control**
  see **emission and exposure control**.

- **exposure investigation**
  the collection and analysis of site-specific information and biological tests (when appropriate) to determine whether people have been exposed to hazardous substances.

- **exposure limit**
  general term defining an administrative substance concentration or intensity of exposure that should not be exceeded. See **discharge limit**.

- **exposure pathway**
  the route a substance takes from its source (where it began) to its end point (where it ends), and how people can come into contact with (or get exposed to) it. An exposure pathway has five parts: a source of contamination (such as an abandoned business), an environmental media and transport mechanism (such as movement through groundwater), a point of exposure (such as a private well), a route of exposure (eating, drinking, breathing, or touching), and a receptor population (people poten-

tially or actually exposed). When all five parts are present, the exposure pathway is termed a completed exposure pathway.

- **exposure ratio**
  in a case control study, value obtained by dividing the rate at which persons in the case group are exposed to the risk factor (or to the protective factor) by the rate at which persons in the control group are exposed to the risk factor (or to the protective factor) of interest.

- **exposure registry**
  a system of ongoing follow up of people who have had documented environmental exposures.

- **exposure test**
  determination of the level, concentration or uptake of a potentially toxic compound and/or its metabolite(s) in biological samples from an organism (blood, urine, hair etc.) and the interpretation of the results to estimate the absorbed dose or degree of environmental pollution, or the measuring of biochemical effects, usually not direct adverse effects of the substance, and relating them to the quantity of substance absorbed, or to its concentration in the environment. See **biological monitoring, biological assessment of exposure**.

- **exposure-dose reconstruction**
  a method of estimating the amount of people's past exposure to hazardous substances. Computer and approximation methods are used when past information is limited, not available, or missing.

- **exposure-effect relationship**
  see **concentration-effect relationship**, .

- **exposure-response relationship**
  see **concentration-response relationship, dose-response relationship**.

- **external validity**
  the extent to which the results of a particular study may be generalised, beyond the limits of the population actually studied. See **validity**.

- **extra risk**
  probability that an agent produces an observed re-

sponse, as distinguished from the probability that the response is caused by a spontaneous event unrelated to the agent. See **risk**.

- **extraneous residue limit (ERL)**

refers to a pesticide residue or contaminant arising from environmental sources (including former agricultural uses) other than the use of a pesticide or contaminant substance directly or indirectly on the commodity. The mass content is expressed in milligrams of pesticide residue or contaminant per kilogram of commodity.

- **extrapolation**

calculation, based on quantitative observations in exposed test species or *in vitro* test systems, of predicted dose-effect and dose-response relationships for a substance in humans and other biota including interspecies extrapolations and extrapolation to susceptible groups of individuals. The term may also be used for qualitative information applied to species or conditions that are different from the ones in which the original investigations were carried out.

- **faecal coliform bacteria**

found in the intestinal tracts of mammals, this bacteria in water or sludge is an indicator of pollution and possible contamination by pathogens.

- **false negative**

a determination that a material is non-toxic when it is in fact toxic.

- **false positive**

a determination of toxicity when the material is in fact non-toxic.

- **fasting glucose test**

a blood test done to determine the plasma glucose concentration in the fasting state (the person has not eaten for 8 to 12 hours). A fasting plasma concentration > 140 mg/dl on at least two occasions is diagnostic for diabetes mellitus.

- **feasibility study**

1. a study to determine the best way to clean up environmental contamination. A number of factors are con-

sidered, including health risk, costs, and what methods will work well.

2. a study that compares different ways to clean up a contaminated site. The feasibility study recommends one or more actions to remediate the site. See **'Remedial investigation'**.

- **fecundity**
ability to produce offspring frequently and in large numbers. See **fertility**.

- **feedstock**
raw material supplied to a machine or processing plant from which other products can be made. For example, polyvinyl chloride and polyethylene are raw chemicals used to produce plastic tiles, mats, fenders, cushions, and traffic cones.

- **fermentation**
chemical reactions accompanied by living microbes that are supplied with nutrients and other critical conditions such as heat, pressure, and light that are specific to the reaction at hand.

- **feromone**
see **pheromone**.

- **ferric chloride**
anhydrous ferric chloride ($FeCl_3$) is a chemical coagulant. Ferric ions are Fe^{3+} and ferrous ions are Fe^{2+}. Both are used in phosphorus precipitation.

- **ferritin**
the primary form of storage iron.

- **fertiliser**
materials such as nitrogen and phosphorus that provide nutrients for plants. Commercially sold fertilisers may contain other chemicals or may be in the form of processed sewage sludge.

- **fertility**
ability to conceive and to produce offspring: for litter-bearing species the number of offspring per litter is used as a measure of fertility. Reduced fertility is sometimes referred to as sub-fertility. See **fecundity**

- **fertility toxicant**
produces abnormalities of male or female reproductive functions or impairs reproductive capacity. See **developmental toxicity**, **reproductive toxicant**.

• **foetotoxicity**
toxicity to the foetus. See **embryotoxicity**, **teratogenicity**.

• **foetus**
young mammal within the uterus of the mother from the visible completion of characteristic organogenesis until birth. In humans, this period is usually defined as from the third month after fertilisation until birth (prior to this, the young mammal is referred to as an embryo). See **embryo**.

• **fibrin**
clot forming protein — these clots often remain suspended in the serum and cause problems within sampling.

• **fibrosis**
abnormal formation of fibrous tissue.

• **fiducial limit**
form of confidence limit given as a stated probability, for example $P = 0.95$. In toxicology the terms fiducial limits and confidence limits are generally considered to be synonymous.

• **filter**
a porous membrane or mat of fibres used to collect particles from the air.

• **fine particle**
particles less than about 2 μm in size, consisting of particles in the nuclei and accumulation modes, term used in describing atmospheric aerosols.

• **first draw**
the water that comes out when a faucet in the kitchen or bathroom is first opened, which is likely to have the highest level of lead contamination from old plumbing solder and pipes.

• **first-order process**
a first order process is a chemical process where the rate is directly proportional to the amount of chemical present. Any process changing at a constant fractional rate.

• **first-pass effect**
bio-transformation of a substance in the liver after absorption from the intestine and before it reaches the systemic circulation.

- **fixed dose procedure**
acute toxicity test in which a substance is tested initially at a small number (3 or 4) predefined doses to identify which produces evident toxicity without lethality. The test may be repeated at one or more higher or lower defined discriminating doses to satisfy the criteria. See **limit test**.

- **flammable**
describes any material that can be ignited easily and that will burn rapidly.

- **flare**
a device that burns gaseous materials to prevent them from being released into the environment. Flares may operate continuously or intermittently and are usually found on top of a stack. Flares also burn off methane gas in a landfill.

- **flash point**
the lowest temperature at which evaporation of a substance produces enough vapour to form an ignitable mixture with air.

- **flocculants**
1. materials which can form gelatinous clouds of precipitate to enclose fine particles of suspended dirt to settle them from the water.
2. those water soluble organic polymers that are used alone or in conjunction with inorganic coagulants to agglomerate solids. The large dense flocks resulting from this process permit rapid and more efficient solids-liquid separations.

- **floodplain**
mostly level land along rivers and streams that may be submerged by floodwater. A 100-year floodplain is an area which can be expected to flood once in every 100 years.

- **flotation**
a process for separating solids from water by developing a froth in a vessel in such fashion that the solids attach to air particles and float to the surface.

- **flow cytometry**
a method of measuring physical and chemical attributes of cells or other biological particles by sensors, as they move by, one after the other, homoge-

neously suspended in fluid. This rapid (about 10,000 particles per second) analytic method allows identification of important cell types, such as malignant cells, T cells and B cells. This process also may utilise immunological binding methods to further identify particular cell groups.

- **flue gas desulphurisation**
the removal of sulphur oxides from exhaust gases of a boiler or industrial process, usually a wet scrubbing operation which concentrates hazardous materials in a slurry, requiring proper disposal.

- **fluorescein**
a molecule that emits light at certain wavelengths, often used to tag antibodies or other proteins and nucleic acids.

- **fluorescence**
the emission of light of one wavelength after being exposed to light of a different wavelength.

- **fluorescence polarisation immunoassay (FPA)**
see **assay**.

- **fluorides**
gaseous, solid, or dissolved compounds containing fluorine that result from industrial processes.

- **fluorocarbons**
any of a number of organic compounds analogous to hydrocarbons in which one or more hydrogen atoms are replaced by fluorine. Once used in the United States as a propellant in aerosols, they are now primarily used in coolants and some industrial processes. FCs containing chlorine are called chlorofluorocarbons (CFCs). They are believed to be modifying the ozone layer in the stratosphere thereby allowing more harmful solar radiation to reach the Earth's surface.

- **fluoroscopy**
a radiologic technique in which a fluoroscope is used to visually examine the body or an organ. (A fluoroscope utilises an X-ray tube and fluorescent screen, with the area to be viewed placed between the screen and the tube.) This immediate imaging, when coupled

with an image intensifier, is invaluable in situations such as cardiac catheterisation, thin needle biopsies of tumours, and localisation of foreign bodies.

- **fluorosis**
an abnormal condition caused by excessive intake of fluorine, characterised chiefly by mottling of the teeth.

- **foci**
the medical term applied to a small group of cells occurring in an organ and distinguishable, either in appearance or histochemically, from the surrounding tissue.

- **follow-up study**
investigation in which individuals or populations, selected on the basis of whether they have been exposed to risk, have received a specified preventive or therapeutic procedure, or possess a certain characteristic, are followed to assess the outcome of exposure, the procedure, or effect of the characteristic, for example, occurrence of disease. See **cohort study**.

- **food additive**
any substance not normally consumed as a food by itself and not normally used as a typical ingredient of the food, whether or not it has nutritive value, the intentional addition of which to food for a technological (including organoleptic) purpose in the manufacture, processing, preparation, treatment, packing, packaging, transport or holding of such food results, or may be reasonably expected to result (directly or indirectly) in it or its by products becoming a component of or otherwise affecting the characteristics of such foods. The term does not include "contaminants" or substances added to food for maintaining or improving nutritional qualities.

- **food allergy**
hypersensitivity reaction to substances in the diet to which an individual has previously been sensitised. See **allergy, hypersensitivity**.

- **food chain**
sequence of transfer of matter and energy in the form

of food from organism to organism in ascending or descending trophic levels.

- **food intolerance**
physiologically based reproducible, unpleasant (adverse) reaction to a specific food or food ingredient that is not immunologically based. See **food allergy**.

- **food web**
network of food chains.

- **forced diuresis**
method of stimulating diuresis based on performing hydrational therapy, sometimes with parallel introduction of diuretics, with the aim of achieving increased clearance of a toxic substance in urine.

- **foreign substance (foreign compound)**
see **xenobiotic**.

- **frame shift mutation**
such a mutation is a change in the structure of DNA causing the transcription of genetic information into RNA to be completely altered because the start point for reading has been changed. In other words, the reading frame for transcription has been altered.

- **fugacity**
broadly speaking, this word is applied to the tendency for a substance to move from one environmental compartment to another. Originally the term was applied to the tendency of a gas to expand or escape and related to its pressure in the system being studied.

- **fugitive emissions**
air pollutants released to the air other than those from stacks or vents, typically small releases from leaks in plant equipment such as valves, pump seals, flanges, sampling connections, etc.

- **fulminant**
sudden, severe or rapid.

- **fume**
an aerosol consisting of small solid particles produced by the condensation of vapours or gaseous combustion products. Usually, such particles are aggregates made up from large numbers of very small primary particles, with the individual units having dimensions of the order of a few

nanometers (nm) and upwards although less than 1μm.

- **fumigant**
substance that is vaporised in order to kill or repel pests.

- **fungi**
(singular, Fungus) moulds, mildews, yeasts, mushrooms, and puffballs, a group of organisms that lack chlorophyll (i.e., are not photosynthetic) and which are usually non-mobile, filamentous, and multicellular. Some grow in the ground, others attach themselves to decaying trees and other plants, getting their nutrition from decomposing organic matter. Some cause disease, others stabilise sewage and break down solid wastes in composting.

- **fungicide**
substance intended to kill fungi.

- **fungus preparation**
substance obtained from fungi that has an insecticidal effect reflecting the pathogenicity of the fungi for insects.

- **gamete**
reproductive cell (either sperm or egg) containing a haploid set of chromosomes. See **zygote**.

- **gametocide**
substance intended to kill gametes.

- **gamma-glutamyltranspeptidase (GGTP)**
an enzyme associated with the uptake of amino acids (protein building blocks) by cells. Elevated blood levels are frequently seen in alcoholic liver disease.

- **garbage**
food waste (animal and vegetable) resulting from the handling, storage, packaging, sale, preparation, cooking, and serving of foods.

- **gastrointestinal**
pertaining to or communicating with the stomach and intestine.

- **gastroschisis**
a congenital defect in which a fissure is present in the wall of the abdomen, often with protrusion.

• **gavage**
administration of materials directly into the stomach by oesophageal intubation.

• **gene**
structurally a basic unit of hereditary material. It is an ordered sequence of nucleotide bases that encodes one polypeptide chain (following transcription to mRNA). See **cistron**.

• **gene amplification**
production of extra copies of a chromosomal sequence found either as intra- or extra-chromosomal DNA. With respect to a plasmid, it refers to the increase in the number of plasmid copies per cell induced by a specific treatment of transformed cells.

• **genetic engineering**
a process of inserting new genetic information into existing cells in order to modify an organism for the purpose of changing particular characteristics.

• **genetic polymorphism**
condition in which a genetic character occurs in more than one form, resulting in the coexistence of more than one morphological type in a given population.

• **genetic toxicity**
damage to DNA with a consequent altered genetic expression. This is also referred to as mutagenesis. The genetic change is a mutation and the causative agent is a mutagen. The change may be heritable (a mutation in a germ cell) or not (a mutation in a somatic cell).

• **genetic toxicology**
study of substances that can produce adverse heritable changes.

• **genome**
the general term for all the genes carried by a cell.

• **genotoxic**
this is the adjective applied to any substance that is able to cause harmful changes to DNA. Such changes may lead to a transformed cell which can form a malignant tumour.

• **genotoxicity**
ability to cause damage to genetic material. Such damage may be mutagenic and/or carcinogenic.

- **genotype**
  genetic constitution of an organism as revealed by genetic or molecular analysis. The complete set of genes, both dominant and recessive possessed by a particular organism, cell, organelle or virus.

- **geographic information system (GIS)**
  a mapping system that uses computers to collect, store, manipulate, analyse, and display data. For example, GIS can show the concentration of a contaminant within a community in relation to points of reference such as streets and homes.

- **geometric standard deviation**
  a measure of dispersion in a lognormal distribution (always greater than or equal to 1.0).

- **germ-free animal**
  animal grown under sterile conditions in the period of postnatal development: such animals are usually obtained by Caesarean operation and kept in special sterile boxes in which there are no viable micro-organisms (sterile air, food and water are supplied). See **axenic animal, gnotobiont, gnotobiote**.

- **germinal aplasia**
  complete failure of gonad development.

- **glaucoma test**
  examination of the eye for the presence of a disease that can result in complete and permanent vision loss and that is characterised by increased pressure within the eyeball and damaged optic disc. When diagnosed early, glaucoma usually can be controlled by the use of suitable eye drops. Continued use of the medication is vital for sight preservation. A test for glaucoma is advised for all adults every three to five years, as the onset of the chronic form of the disease may be without noticeable symptoms.

- **global warming**
  see **climate change**.

- **glomerular**
  pertaining to a tuft or cluster, as of a plexus of capillary blood vessels or nerve fibres, especially referring to the capillaries of the glomeruli of the kidney.

- **glucagon**
a polypeptide hormone secreted by the alpha cells of the islets of Langerhans in response to hypoglycaemia. Can be given as a intramuscular injection to treat severe hypoglycaemia.

- **glucose**
the main sugar and source of energy in blood. It is contained in common dietary sugar (sucrose). Also known as dextrose, it occurs widely in nature, especially in certain plants. Blood and urine levels of glucose can help diagnose diabetes, hypoglycemia, and hormone disorders. Glucose tests are often called tests for 'sugar'. A glucose tolerance test, usually performed to assist in the diagnosis of diabetes or hypoglycemia, measures the ability of the body to metabolise a standard amount of the sugar.

- **glucose test**
a blood test done to determine the non-fasting plasma glucose concentration. Levels of > 200 mg/dl plus classic signs and symptoms of diabetes including polydipsia, polyuria, polyphagia and weight loss are diagnostic for diabetes mellitus.

- **glycosylated haemoglobin**
the iron-containing red pigment of red blood cells (haemoglobin) combined with a glucose derivative. When glucose is elevated, more glycosylate product is formed in the cells. Measurement of this haemoglobin reflects the blood glucose level over the previous several weeks. Thus, it is a good indicator of blood glucose control in diabetics.

- **gnotobiont**
see **gnotobiote, germ-free animal**.

- **gnotobiote**
specially reared laboratory animal whose microflora and microfauna are specifically known in their entirety. See **germ-free animal**.

- **goiter**
a swelling of the thyroid gland, possible symptom of hypo- or hyperthyroidism.

- **gonadotropic**
pertaining to effects on sex glands and on the systems that regulate them.

- **gonorrhea**
a chiefly sexually transmitted infection caused by gonococci bacteria. If left untreated, gonorrhea can result in complications including sterility, blindness (preventive treatment is thus routinely given to newborns), or it spreads to the heart and joints. Women with gonorrhea may be asymptomatic, whereas men often present with a noticeable discharge from the penis.

- **good agricultural practice (GAP) in the use of pesticides**
nationally authorised safe uses of pesticides under actual conditions necessary for effective and reliable pest control. It encompasses a range of levels of pesticide applications up to the highest authorised use, applied in a manner that leaves a residue which is the smallest amount practicable. Authorised safe uses include nationally registered or recommended uses, that take into account public and occupational health and environmental safety considerations. Actual conditions include any stage in the production, storage, transport, distribution, and processing of food commodities and animal feed.

- **good laboratory practice (GLP) principles**
fundamental rules incorporated in national regulations concerned with the process of effective organisation and the conditions under which laboratory studies are properly planned, performed, monitored, recorded, and reported. See **quality assurance, quality control**.

- **good manufacturing practice (GMP) principles**
fundamental rules incorporated in national regulations concerned with the process of effective organization of production and ensuring standards of defined quality at all stages of production, distribution and marketing. Minimisation of

waste and its proper disposal are part of this process.

- **graded effect**
  result that can be measured on a graded scale of intensity or severity and its magnitude related directly to the dose or concentration of the substance producing it. See **all-or-none effect, quantal effect, stochastic effect.**

- **gradient**
  the change in a property over a certain distance. For example, lead can accumulate in surface soil near a road due to automobile exhaust. As you move away from the road, the amount of lead in the surface soil decreases. This change in the lead concentration with distance from the road is called a gradient.

- **grand rounds**
  training sessions for physicians and other health care providers about health topics.

- **granuloma**
  granular growth or tumour, usually of lymphoid and epithelial cells.

- **Grave's disease**
  a disease characterised by excess thyroid hormone production, caused by a stimulation of the thyroid gland by antibodies to the TSH receptor. Ophthalmopathy may or may not be present.

- **greenhouse effect**
  see **climate change.**

- **ground treatment of plants**
  dusting or spraying of plants with pesticides by hand, by special machines, or by apparatus fixed to tractors or driven by them.

- **ground water**
  water found below the surface of the land, usually in porous rock formations. Ground water is the source of water found in wells and springs and is used frequently for drinking.

- **guideline value**
  quantitative measure (a concentration or a number) of a constituent of an environmental medium that ensures aesthetically pleasing air, water, or food and does not result in a significant risk to the user.

- **guides to air quality**
  sets of atmospheric concentrations and exposure times that are associated with specific effects of varying degrees of pollution on man, animals, vegetation, and the environment in general.

- **guides to environmental quality**
  sets of concentrations, numbers and exposure times that are associated with the specific effects of factors in environmental media on man, animals, vegetation, and the environment in general.

- **guinea pig maximisation test**
  this is a skin test for screening possible contact allergens. It is considered to be a useful model for predicting likely moderate and strong sensitisers in humans.

- **habitat**
  the sum of environmental conditions in a specific place that is occupied by an organism, population or community.

- **haematemesis**
  vomitting of blood.

- **haematoma**
  localised accumulation of blood, usually clotted, in an organ, space, or tissue, due to a failure of the wall of a blood vessel.

- **haematuria**
  presence of blood in the urine.

- **haemodialysis**
  use of an artificial kidney to remove toxic compounds from the blood by passing it through a tube of semi-permeable membrane. The tube is bathed in a dialysing solution to restore the normal chemical composition of the blood while permitting diffusion of toxic substances from the blood.

- **haemoglobinuria**
  presence of free haemoglobin in the urine.

- **haemolysin**
  substance that damages the membrane of erythrocytes causing the release of haemoglobin.

- **haemolysis**
  release of haemoglobin from erythrocytes, and its appearance in the plasma.

- **haemoperfusion**
  passing blood through a column of charcoal or adsorbent resin for the removal of drugs or toxins.

- **haemosiderin**
  iron-containing pigment that is formed from haemoglobin released during the disintegration of red blood cells and that accumulates in individuals who have ingested excess iron.

- **haemosiderin**
  this is the name given to the iron-protein molecule which is a source of iron for haemoglobin synthesis and other processes requiring iron.

- **half-life**
  1. the time taken by certain materials to lose half their strength. For example, the half life of DDT is 15 years, of radium 1,580 years. The time required for half of the atoms of a radioactive element to undergo decay.
  2. the time required for the elimination of one half a total dose from the body.

- **halogen**
  any of a group of 5 chemically-related non-metallic elements that includes bromine, fluorine, chlorine, iodine, and astatine.

- **halon**
  bromine containing compounds with long atmospheric life-times whose breakdown in the stratosphere causes depletion of ozone. Halons are used in fire fighting.

- **haploid**
  this adjective is applied to a cell containing only one set of chromosomes.

- **haploid (monoploid)**
  state in which a cell contains only one set of chromosomes. See **diploid, gamete, meiosis**.

- **hapten**
  low-molecular-weight molecule that contains an antigenic determinant (epitope) that may bind to a specific antibody but which is not itself antigenic unless bound with an antigenic carrier such as a protein or cell. Once bound it can cause the sensitisation of lymphocytes, possibly leading to allergy or cell-mediated hypersensitivity. See **allergy, antigen, antibody, cell-me-**

diated hypersensitivity, epitope.

- **harm**
damage or adverse effect to a population, species, individual organism, organ, tissue or cell. See **adverse effect.**

- **harmful occupational factor**
component of the work environment the effect of which on a worker under certain conditions leads to ill-health or reduction of working ability.

- **harmful substance**
substance that can cause ill-health or adverse effects either at the time of exposure or later in the life of the present and future generations. See **noxious substance.**

- **Hashimoto's thyroiditis**
an autoimmune destructive disease characterised by atrophy of the thyroid gland.

- **hazard**
a source of potential harm from past, current, or future exposures.

- **hazard assessment**
determination of factors controlling the likely effects of a hazard such as the dose-effect and dose-response relationships, variations in target susceptibility, and mechanism of toxicity. See **exposure assessment, hazard evaluation, hazard identification, risk assessment, risk characterisation, risk estimation, risk evaluation, risk identification, risk perception.**

- **hazard evaluation**
establishment of a qualitative or quantitative relationship between hazard and benefit, involving the complex process of determining the significance of the identified hazard and balancing this against identifiable benefit: this may subsequently be developed into a risk evaluation. See **exposure evaluation, hazard assessment, hazard identification, risk assessment, risk characterisation, risk estimation, risk evaluation, risk identification, risk perception.**

- **hazard identification**
determination of substances of concern, their adverse effects, target populations,

and conditions of exposure, taking into account toxicity data and knowledge of effects on human health, other organisms and their environment.

- **hazard quotient (HQ)**
ratio of toxicant exposure (estimated or measured) to a reference value regarded as corresponding to a threshold of toxicity. If the total hazard quotient from all toxicants to a target exceeds unity, the combination of toxicants may produce (will produce under assumptions of additivity) an adverse effect. See **hazard, pollutant, toxic substance.**

- **hazardous production factor**
production factor the effect of which on a worker under certain conditions results in injury or some impairment of health.

- **hazardous waste**
potentially harmful substances that have been released or discarded into the environment.

- **health**
state of complete physical, mental and social well-being, and not merely the absence of disease or infirmity.

- **health assessment for contaminated sites**
determination of actual or possible health effects due to environmental contamination or exposure. It includes a health-based interpretation of all the information known about the situation. The information may come from site investigations (environmental sampling and studies), exposure assessments, risk assessments, biological monitoring or health effects studies. The health assessment is used to advise people how to prevent or reduce their exposures, to determine remedial actions or the need for additional studies.

- **health consultation**
a review of available information or collection of new data to respond to a specific health question or request for information about a potential environmental hazard. Health consultations are focussed on a specific exposure issue. Health consultations are, therefore,

more limited than a public health assessment, which reviews the exposure potential of each pathway and chemical.

- **health effects studies related to contaminants**
  studies of the health of people who may have been exposed to contaminants. They include, but are not limited to, epidemiological studies, reviews of health status of people in exposure or disease registries, and doing medical tests.

- **health hazard**
  any factor or exposure that may adversely affect health.

- **health investigation**
  the collection and evaluation of information about the health of community residents. This information is used to describe or count the occurrence of a disease, symptom, or clinical measure and to evaluate the possible association between the occurrence and exposure to hazardous substances.

- **health promotion**
  the process of enabling people to increase control over, and to improve, their health.

- **health registry**
  a record of people exposed to a specific substance (such as a heavy metal), or having a specific health condition (such as cancer or a communicable disease).

- **health statistics review**
  the analysis of existing health information (i.e., from death certificates, birth defects registries, and cancer registries) to determine if there is excess disease in a specific population, geographic area, and time period. A health statistics review is a descriptive epideimologic study.

- **health surveillance**
  periodic medico-physiological examinations of exposed workers with the objective of protecting health and preventing occupationally related disease. See **biological monitoring**, **biomarker**, **monitoring**.

- **health-based exposure limit**
  maximum concentration or intensity of exposure that can be tolerated without significant effect (based on only

scientific and not economic evidence concerning exposure levels and associated health effects).

- **healthy worker effect**
epidemiological phenomenon observed initially in studies of occupational diseases. Workers usually exhibit lower overall disease and death rates than the general population, due to the fact that the old, severely ill and disabled are ordinarily excluded from employment. Death rates in the general population may be inappropriate for comparison, if this effect is not taken into account.

- **heavy metal**
a common hazardous waste. It can damage organisms at low concentrations and tends to accumulate in the food chain.

- **Henry's law constant**
at constant temperature and pressure, the ratio of the partial pressure of a gas above a liquid to its molal solubility in the liquid and therefore a measure of its partition between the gas phase and the solute phase.

- **hepatic**
pertaining to the liver.

- **hepatitis**
inflammation of the liver, usually caused by bacterial or viral infection, drugs (including alcohol), toxins, or parasites. This disease is characterised by an enlarged liver, jaundice, diminished appetite, nausea, and abdominal pain. There are at least three forms of viral hepatitis (hepatitis A, hepatitis B, and non-A, non-B hepatitis). Hepatitis A, also known as infectious hepatitis, is transmitted by contaminated food or water and personal contact, and is most often found in children or young adults. Hepatitis B, formerly called serum hepatitis, may be transmitted through contaminated blood or needles, as well as sexually, and has a longer incubation period than that of hepatitis A. Hepatitis B can result in chronic active hepatitis. Non-A, non-B hepatitis is also transmitted through blood and can cause severe liver damage. Radioimmunoassays and enzyme-linked immunosorbent assays may

be used to screen for hepatitis.

- **hepatitis B surface antigen (HBSAG)**
  a part of the hepatitis B virus that, when in the blood, is one of the earliest markers of infection with hepatitis B virus, appearing even before symptoms. Also found in blood of a carrier. HBSAG is tested for in the differential diagnosis of hepatitis.

- **hepatocyte**
  in histology, this name is given to a parenchymal liver cell.

- **hepatoma**
  a malignant tumour of the liver (primary hepatocellular carcinoma) sometimes associated with hepatitis or cirrhosis. Elevated blood levels of alpha-foetoprotein (AFP) may indicate the presence of a hepatoma. AFP levels are useful in monitoring chemotherapeutic success in treating this condition.

- **hepatotoxic**
  poisonous to liver cells.

- **herbicide**
  substance intended to kill plants.

- **heterotrophic organism**
  consumers such as humans and animals, and decomposers- chiefly bacteria and fungi- that are dependent on organic matter for food.

- **high density lipoprotein cholesterol (HDLC)**
  the fraction of plasma lipoprotein cholesterol component with most density (relative weight). It is thought to affect blood cholesterol levels by removing it from plasma and tissues and carrying it back to the liver for excretion after being acted on by bile. High levels of HDLC are considered to place an individual at a lower risk for coronary heart disease (atherosclerosis). Low levels of HDLC are associated with an increased risk for the disease. The density may be established by ultracentrifugation or electrophoresis.

- **hindered settling**
  a stage of settling where the accumulated settled solids have compacted to an ex-

tent that egress of water from the mass in hindered and, therefore, settling is slowed.

- **histochemistry**
this is the science concerned with the chemistry of microscopically visible structures.

- **histogenic origin**
germ cell layer of the embryo from which a given adult tissue develops.

- **histogram**
a graphic means of comparing magnitudes of frequencies or numbers of items; usually shown in bar graphs or columns.

- **histology**
study (usually microscopic) of the anatomy of tissues and their cellular and subcellular structure.

- **histopathology**
microscopic pathological study of the anatomy and cell structure of tissues in disease to reveal abnormal or adverse structural changes.

- **homeostasis**
normal, internal stability in an organism maintained by co-ordinated responses of the organ systems that automatically compensate for environmental changes.

- **homocysteine**
homocysteine is an amino acid derived from the digestion of protein-rich foods.

- **homology**
degree of identity existing between the nucleotide sequences of two related but not complementary DNA or RNA molecules. 70 % homology means that on the average 70 out of every 100 nucleotides are identical in a given sequence. The same term is used in comparing the amino acid sequences of related proteins.

- **hormesis**
stimulatory effect of small doses of a potentially toxic substance that is inhibitory in larger doses.

- **hormone**
chemical substances secreted by organs or parts of the body, especially the endocrine glands, into the bloodstream. Each hormone has a specific regulatory or functional effect.

- **household or domestic waste**
  solid waste, composed of garbage and rubbish, which normally originates from residential, private households, or apartment buildings. Domestic waste may contain a significant amount of toxic or hazardous waste from improperly discarded pesticides, paints, batteries, and cleaners.

- **human ecology**
  interrelationship between humans and the entire environment - physical, biological, socio-economic, and cultural, including the interrelationships between individual humans or groups of humans and other human groups or groups of other species.

- **human equivalent dose**
  human dose of an agent that is believed to induce the same magnitude of a toxic effect that the known animal dose has induced.

- **hydraulic grade line**
  the surface or profile of water flowing off hydraulic gradient The slope of the hydraulic grade line is under pressure, the hydraulic grade line is at the level water would rise to in a small vertical tube connected to the pipe. Also see **energy grade line**

- **hydraulic gradient**
  the direction of ground water flow due to changes in the depth of the water table.

- **hydrocarbons**
  chemicals that consist entirely of hydrogen and carbon. Hydrocarbons contribute to air pollution problems like smog.

- **hydrochlorination**
  the application of hypochlorite compounds to water for the purpose of disinfection.

- **hydrochlorinators (hi-poe-klor-uh-nay-tors)**
  chlorine pumps, chemical feed pumps or devices used to dispense chlorine solutions made from hypochlorites such as bleach (sodium hypochlorite) or calcium hypochlorite into the water being treated.

- **hydrogen sulphide (H2S)**
  the gas emitted during organic anaerobic decomposition that smells like rot-

ten eggs. It is also a by-product of oil refining and burning and can cause illness in heavy concentrations.

- **hydrogeologic conditions**
conditions stemming from the interaction of ground water and the surrounding soil and rock.

- **hydrogeologic cycle**
the natural process recycling water from the atmosphere down to (and through) the earth and back to the atmosphere again.

- **hydrogeologist**
a person who studies and works with groundwater.

- **hydrogeology**
the geology of ground water, with particular emphasis on the chemistry and movement of water.

- **hydrograph**
a graph of the rate of runoff plotted against time for a point on a channel.

- **hydrolysis**
1. a chemical reaction in which a compound reacts with water, resulting in its decomposition or alteration.
2. the splitting of a salt and water into its ions and formation of a weak acid or base, or both.

- **hydrophilic**
1. having a strong affinity (liking) for water. The opposite of hydrophobic.
2. describing the character of a molecule or atomic group which has an affinity for water.

- **hydrophobic**
having a strong aversion (dislike) for water. The opposite of hydrophilic.

- **hydropneumatic**
a water system, usually small, in which a water pump is automatically controlled (started and stopped) by the air pressure in a compressed-air tank.

- **hydrosphere**
this is the broad name for the water above, on or in the earth's crust, including oceans, seas, lakes, ground water, and atmospheric moisture.

- **hydrostatic pressure**
1. the pressure at a specific elevation exerted by a body of water at rest.

2. in the case of groundwater, the pressure at a specific elevation due to the weight of water at higher levels in the same zone of saturation.

- **hydroxyl**
  the anion of water or OH^-, also present in all hydroxides.

- **hygiene**
  science of health and its preservation.

- **hyper-**
  prefix meaning above or excessive. When used with the suffix '-aemia' refers to blood and with the suffix '-uria' refers to urine, for e.g. 'hyperbilirubinaemia'.

- **hyperaemia**
  excessive amount of blood in any part of the body.

- **hyperalimentation**
  ingestion or administration of nutrients in excess of optimal amounts.

- **hyperbilirubinaemia**
  excessive concentration of bilirubin in the blood.

- **hypercalcaemia**
  excessive concentration of calcium in the blood.

- **hyperglycaemia**
  excessive concentration of glucose in the blood.

- **hyperkalaemia**
  excessive concentration of potassium in the blood.

- **hypernatraemia**
  excessive concentration of sodium in the blood.

- **hyperparathyroidism**
  abnormally increased parathyroid gland activity that affects, and is affected by, plasma calcium concentration.

- **hyperplasia**
  abnormal multiplication or increase in the number of normal cells in a tissue or organ. See **hypertrophy, neoplasia**.

- **hyper-reactivity**
  term used to describe the responses of (effects on) an individual to (of) an agent when they are qualitatively those expected, but quantitatively increased.

- **hyper-reflexia**
  exaggeration of reflexes.

- **hypersensitivity**
  state in which an individual reacts with allergic effects

following exposure to a certain substance (allergen) after having been exposed previously to the same substance. See **allergy**, **cell-mediated hypersensitivity**, **sensitisation**.

- **hypersusceptibility**
excessive reaction following exposure to a given amount or concentration of a substance as compared with the large majority of other exposed subjects. See **idiosyncrasy**.

- **hypertension**
persistently high blood pressure in the arteries or in a circuit, for example pulmonary hypertension or hepatic portal hypertension.

- **hyperthyroidism**
a condition characterised by accelerated metabolism caused by an excessive amount of thyroid hormones (overactive thyroid).

- **hypervitaminosis**
condition resulting from the ingestion of an excess of one or more vitamins.

- **hypo-**
prefix meaning under, deficient. When used with the suffix -aemia' refers to blood and with the suffix '-uria' refers to urine, for example 'hypocalcaemia'.

- **hypocalcaemia**
abnormally low calcium concentration in the blood.

- **hypochlorite**
chemical compounds containing available chlorine used for disinfection. They are available as liquids (bleach) or solids (powder, granules and pellets). Salts of hypochlorous acid.

- **hypoglycemia (low blood sugar)**
a condition in which there is an abnormally diminished concentration of glucose in the blood, which may lead to convulsions or coma.

- **hypokalaemia**
abnormally low potassium concentration in the blood.

- **hypolimnion**
the lowest layer in a thermally stratified lake or reservoir. This layer consists of colder, more dense water, has a constant temperature and no mixing occurs.

- **hyponatraemia**
abnormally low sodium concentration in the blood.

- **hypothalamus**
an area of the brain that regulates endocrine activity as well as somatic functions e.g.body temperature, sleep, appetite. Hypothalamic neurohormones (e.g.TRH) control various functions of the pituitary gland.

- **hypothesis testing**
a statistical technique (e.g. Dunnettis test) for determining whether a tested concentration is statistically different from the control. End-points determined from hypothesis testing are NOEC and LOEC. The two hypotheses commonly tested in WET are Null Hypothesis (Ho) The effluent is not toxic. Alternate hypothesis (Ha) The effluent is toxic.

- **hypothyroidism**
a condition characterised by deficient production of thyroid hormones (underactive thyroid).

- **hypotriglyceridaemia**
in medicine, this defines the situation characterised by a decreased blood triglyceride content.

- **hypovolaemic**
pertaining to an abnormally decreased volume of circulating fluid (plasma) in the body.

- **hypoxaemia**
deficient oxygenation of the blood.

- **hypoxia**
abnormally low oxygen content or tension. Deficiency of oxygen in the inspired air, in blood or in tissues, short of anoxia.

- **i.v.**
this is the abbreviation for intravenous (used in relation to administration of drugs or other substances).

- **iatrogenic**
any adverse condition resulting from medical treatment. See **nosocomial**

- **icterus**
excess of bile pigment in the blood and consequent deposition and retention of bile pigment in the skin and the sclera. See **hyperbilirubinaemia, jaundice**.

- **idiosyncrasy**
  genetically based unusually high sensitivity of an organism to the effect of certain substances. See **hypersusceptibility, pharmacogenetics**.

- **imhoff cone**
  a clear, cone-shaped container marked with graduations. The cone is used to measure the volume of settleable solids in a specific volume (usually one litre) of water.

- **immediately dangerous to life or health concentration (IDLH)**
  this is a regulatory value defined as the maximum exposure concentration in the workplace from which one could escape within 30 minutes without any escape-impairing symptoms or any irreversible health effects. This value should be referred to in respirator selection.

- **immission**
  environmental concentration of a pollutant resulting from a combination of emissions and dispersals (often synonymous with exposure).

- **immune complex**
  product of an antigen-antibody reaction that may also contain components of the complement system.

- **immune response**
  selective reaction of the body to substances that are foreign to it, or that the immune system identifies as foreign, shown by the production of antibodies and antibody-bearing cells or by a cell-mediated hypersensitivity reaction. See **antibody, autoimmune disease, cell-mediated hypersensitivity**.

- **immunochemical**
  describing the chemistry of antigens, antibodies, and their relation to each other.

- **immunochemistry**
  study of biochemical and molecular aspects of immunology, especially the nature of antibodies, antigens and their interactions.

- **immunoelectrophoresis**
  pertaining to investigation of substances and their identifications by a method combining mobility in an electric field and utilising an antigen-antibody reaction. It

refers to a combination technique used to elevate proteins in a mixture, or the characteristics of a specific protein.

- **immunofluorescence**
a method of locating antigens in tissue, or of identifying an antigen by the fluorescence resulting when the antigenic material is exposed to a specific antibody coupled with a fluorescent dye. The antigen-antibody combination then is visible (luminous) using ultraviolet light. This technique is used in the diagnosis of syphilis, toxoplasmosis (a serious disease which may lead to blindness and brain defects in the congenital form, as well as myocarditis and other conditions), and other diseases.

- **immunogen**
see **antigen**.

- **immunoglobulin**
family of closely related glycoproteins capable of acting as antibodies and present in plasma and tissue fluids. Immunoglobulin E is the source of antibody in many hypersensitivity (allergic) reactions. See **allergy**, **antibody**, **hypersensitivity**.

- **immunoglobulin E-mediated hypersensitivity**
state in which an individual reacts with allergic effects caused fundamentally by the reaction of antigen-specific immunoglobulin E following exposure to a certain substance (allergen) after having been exposed previously to the same substance.
See **allergy**, , **antigen**, **cell-mediated hypersensitivity**, **hypersensitivity**, **immunoglobulin**.

- **immunoglobulins (Ig)**
a family of similar but distinctly different proteins which act as antibodies of five major types (IgA, IgD, IgE, IgG, and IgM). They are produced in lymphoid tissue by B lymphocytes and plasma cells, in response to the presence of antigenic substances.

- **immunology**
the study of immunity , the security or protection against a foreign organism or poison introduced into an individual.

- **immunopotentiation**
enhancement of the capacity of the immune system to produce an effective response.

- **immunosorbence**
a method of (or preparation for) using an insoluble substance combined with specific antibody or antigen for absorbing antigen or antibody components from blood. This allows determination of the amount of antigen or antibody in the specimen.

- **immunosuppression**
in medicine, this term is applied to inhibition of the normal response of the immune system to an antigen.

- **immunosurveillance**
mechanisms by which the immune system is able to recognise and destroy malignant cells before the formation of an overt tumour.

- **immunotoxic**
poisonous to the immune system.

- **impactor**
a device in which aerosol particles with sufficiently high inertia in a deflected air stream are impacted onto a surface.

- **impeller**
a rotating set of vanes in a pump designed to pump or lift water.

- **impermeable**
not easily penetrated. The property of a material or soil that des not allow, or allows only with great difficulty, the movement or passage of water.

- **impinger**
a device in which particles are removed by impacting the aerosol particles into a liquid.

- **impotence**
the inability to have an erection.

- **impoundment**
a body of water or sludge confined by a dam, dike, floodgate, or other barrier.

- **in situ**
in place, the original location, in the natural environment.

- **in vitro**
in glass, referring to a study in the laboratory usually involving isolated organ, tis-

sue, cell, or biochemical systems. See **in vivo**.

- **in vivo**
in the living body, referring to a study performed on a living organism. See **in vitro**.

- **incidence**
the number of new cases of disease in a defined population over a specific time period.

- **incidence rate**
measure of the frequency with which new events occur in a population. Value obtained by dividing the number of new events that occur in a defined period by the population at risk of experiencing the event during this period, sometimes expressed as person-time.

- **incineration**
the destruction of solid, liquid, or gaseous wastes by controlled burning at high temperatures. Hazardous organic compounds are converted to ash, carbon dioxide, and water. Burning destroys organics, reduces the volume of waste, and vapourises water and other liquids the wastes may contain. The residue ash produced may contain some hazardous material, such as non-combustible heavy metals, concentrated from the original waste.

- **incinerator**
a furnace for the routine burning of waste materials using controlled flame combustion.

- **incompatible waste**
a waste unsuitable for mixing with another waste or material because of reactivity hazards.

- **incontinence**
a loss of urinary control.

- **incremental unit risk estimate**
for an air pollutant, this is the additional lifetime cancer risk occurring in a hypothetical population in which all individuals are exposed continuously from birth throughout their lifetimes to a concentration of 1 microgram per cubic metre (mg/m^3) of the pollutant in the air they breathe.

- **indicator (chemical)**
a substance that gives a visible change, usually of colour at a desired point in

a chemical reaction, generally at a specified end point.

- **indicator (instrument)**
a device which indicates the result of a measurement. Most indicators in the water utility field use either a fixed scale and movable indicator (pointer) such as a pressure guage or a movable scale and movable indicator like those used on a circular-flow recording chart. Also called a Receiver.

- **indigenous species**
a species that is likely, due to historical presence, to occur at a specified site for some portion of its life span

- **indirect discharge**
the introduction of pollutants from a non-domestic source into the sewage treatment system. Indirect discharges can be commercial or industrial facilities who must pre-treat their wastes before discharge into local sewers.

- **indirect exposure**
1. exposure to a substance in a medium or vehicle other than the one originally receiving the substance.
2. exposure of people to a substance by contact with a person directly exposed. See **bystander exposure, para-occupational exposure**.

- **individual monitor**
see **personal sampler**.

- **individual protective device (IPD)**
device for individual use for protection of the whole body, eyes, respiratory pathways or skin of workers against hazardous and harmful production factors. See **personal protective device (PPD), personal protective equipment (PPE)**.

- **individual risk**
probability that an individual person will experience an adverse effect.

- **indoor air**
breathing air inside a habitable structure, often highly polluted because of lack of exchange with fresh oxygen from outdoors. Solvents, smoke, paints, furniture glues, carpet padding, and other synthetic chemicals trapped inside contribute to an often unhealthy environment.

- **inducer**
  substance that causes induction. See **induction**.

- **induction**
  increase in the rate of synthesis of an enzyme in response to the action of an inducer or environmental conditions: often the substrate of the induced enzyme or a structurally similar substance (gratuitous inducer) that is not metabolised.

- **induction period**
  time from the onset of exposure to the appearance of signs of disease. See **latent period**.

- **industrial waste**
  unwanted materials produced in or eliminated from an industrial operation and categorised under a variety of headings, such as liquid wastes, sludge, solid wastes, and hazardous wastes.

- **inert ingredients**
  substances that are not active, such as water, petroleum distillates, talc, corn meal, or soaps. When discussing pesticides, inert ingredients do not attack a particular pest, but some are chemically or biologically active, causing health and environmental problems.

- **infectious waste**
  see **medical waste.**

- **infiltration**
  1. the gradual flow or movement of water into and through (to percolate or pass through) the pores of the soil. Also see **percolation**.
  2. the penetration of water from the soil into sewer or other pipes through defective joints, connections or manhole walls.

- **infiltration gallery**
  a sub-surface groundwater collection system, typically shallow in depth, constructed with open-jointed or perforated pipes that discharge collected water into a water-tight chamber. From this chamber the water is pumped to treatment facilities and into the distribution system. Infiltration galleries are usually located close to streams or ponds and may be under the direct influence of surface water.

- **infiltration rate**
  quantity of water (usually measured in inches) that will enter a particular type of soil per unit time (usually one hour).

- **influent**
  water or other liquid- raw or partially flowing into a reservoir, basin, treatment process or treatment plant.

- **ingestion**
  the act of swallowing something through eating, drinking, or mouthing objects. A hazardous substance can enter the body this way [see **route of exposure**].

- **inhalable particles**
  those materials that are deposited anywhere in the respiratory tract.

- **inhalation**
  act of drawing in of air, vapour or gas and any suspended particulates into the lung.

- **inherently bio-degradable**
  class of compounds for which there is unequivocal evidence of bio-degradation (primary or ultimate) in any test of bio-degradability. See **bio-degradation**.

- **inhibition concentration (IC)**
  a point estimate of the toxicant concentration that would cause a given percent reduction in a non-lethal biological measurement (e.g., reproduction or growth), calculated from a continuous model (e.g., interpolation method). IC25 is a point estimate of the toxicant concentration that would cause a 25-percent reduction in a non-lethal biological measurement (EPA, 2000). The lower the IC25, the more toxic the chemical or effluent sample. Other IC values, e.g. the IC50 or IC10 may also be calculated to determine concentrations causing a more or less severe effect.

- **inhibitory dose (ID)**
  dose of a substance that causes a defined inhibition of a given system: ID50 is the median dose that causes 50 % inhibition. See **effective dose, lethal dose**.

- **initial compliance period**
  the first full three-year compliance period which begins at least 18 months after promulgation.

- **initiator**

1. agent that induces a change in a chromosome or gene that leads to the induction of tumours after a second agent, called a promoter, is administered to the tissue. See **promoter**.

2. substance that starts a chain reaction. An initiator is consumed in a reaction, in contrast to a catalyst.

- **in-line filtration**

the addition of chemical coagulants directly to the filter inlet pipe. The chemicals are mixed by the flowing water. Flocculation and sedimentation facilities are eliminated. This pretreatment method is commonly used in pressure filter installations. Also see **conventional filtration and direct filtration**.

- **innovative technology**

new or inventive methods to treat hazardous wastes, conserve energy, or prevent pollution.

- **innoculum**

1. Bacteria placed in compost to start biological action.

2. a medium containing organisms which is introduced into cultures or living organisms.

- **inorganic**

1. compounds not containing carbon. Non-plant or animal.

2. material such as sand, salt, iron, calcium salts and other mineral materials. Inorganic substances are of mineral origin, whereas organic substances are usually of animal or plant origin. Also see **organic**.

- **inorganic materials**

compounds derived from other than vegetable or animal sources that do not generally contain carbon atoms. Some simple carbon compounds are considered inorganic (i.e., carbides, CO_x, carbonates, CS_2).

- **inorganic toxic agent**

toxic agents that are elements or inorganic compounds

- **input horsepower**

the total power used in operating a pump and motor. Input HP = (Brake HP)(100%)Motor Efficiency, %)

- **insecticide**
substance intended to kill insects.

- **instream uses**
water uses that can be carried out without removing the water from its source, as in navigation and recreation.

- **instream waste concentration (IWC)**
the concentration of a toxicant in the receiving water after mixing. The IWC is the inverse of the dilution factor. It is sometimes referred to as the receiving water concentration (RWC). This value is frequently used to determine permit limits for toxicity (e.g., no chronic toxicity at the IWC). For effluents the IWC is typically calculated using critical flow conditions of the receiving stream (e.g., 7-day, 10-year low flow or 7Q10) and the design (maximum) flow of the plant effluent IWC = 100 x effluent flow/(effluent flow + 7Q10).

- **insulin**
a protein hormone secreted by the beta cells of the pancreas that signals the fed state. It is secreted in response to elevated blood levels of glucose.

- **insulin resistance**
a defect present in the majority of individuals with impaired glucose tolerance and essentially in all patients with Type 2 diabetes and is an impairment of insulin action.

- **intake**
amount of a substance that is taken into the body, regardless of whether or not it is absorbed. The total daily intake is the sum of the daily intake by an individual from food, drinking-water, and inhaled air.

- **integral indicator of toxic effect**
parameter (such as body weight or temperature) characterising the overall changes in the general state of the organism exposed to a toxic substance.

- **integrated exposure assessment**
a summation over time, in all media, of the magnitude of exposure to a toxic chemical.

- **integrated pest management (IPM)**
  a combination of biological, cultural, and genetic pest control methods with use of pesticides as the last resort. IPM considers a targetted species' life cycle and intervenes in reproduction, growth, or development to reduce the population. Land use practices are examined for possible change, other animals, birds, or reptiles in the ecosystem are used as natural predators.

- **integrator**
  a device or meter that continuously measures and calculates (adds) total flows in gallons, or million cubic feet. or some other unit of volume measurement. Also called a Totaliser

- **interface**
  the common boundary layer between two substances such as water and a solid (metal), or between two fluids such as water and a gas (air), or between a liquid (water) and another liquid (oil).

- **interfacial layer**
  region of space comprising and adjoining the phase boundary within which properties of matter are significantly different from the values in the adjoining bulk phases. See **surface layer**.

- **interflow**
  lateral movement of water in the upper layer of soil.

- **interim remedial measure (IRM)**
  an action taken at a contaminated site to reduce the chances of human or environmental exposure to site contaminants. Interim remedial measures are planned and carried out before comprehensive remedial studies. They can prevent additional damage during the study phase, but don't interfere in any way with the need to develop a complete remedial programme. An example of an interim remedial measure is removing drums of chemicals to a storage facility from a site that has drums sitting in an empty field.

- **interlock**
  an electrical switch, usually magnetically operated. Used to interrupt all (local) power to a panel or device when the door is opened or the circuit exposed to service.

- **intermediate duration exposure**
  contact with a substance that occurs for more than 14 days and less than a year.

- **internal validity**
  selection and comparison of index and comparison groups in such a manner that, apart from sampling error, the observed differences between these groups with respect to dependent variables under study may be attributed only to the hypothesised effect under investigation.

- **interpretation (of data or findings)**
  evaluation of the observations from an investigation or study in order to determine their significance for human health, for the environment or for both.

- **interspecies dose conversion**
  process of extrapolating from the doses of one animal species to another, for example, from rodent dose to human equivalent.

- **interspecies extrapolation model**
  model used to extrapolate from results observed in laboratory animals to humans.

- **interstate carrier**
  any vehicle or transport which conveys passengers in interstate commerce.

- **interstice**
  a very small open space in a rock or granular material. Also called a void or void space. Also **see pore**.

- **interstitial pneumonia**
  chronic form of pneumonia involving increase of the interstitial tissue and decrease of the functional lung tissue.

- **interstitial water**
  water occupying space between sediment or soil particles (syn. Pore water)

- **intervention study**
  epidemiological investigation designed to test a hypothesised cause-effect relationship by modifying a supposed causal factor in a population.

- **intestinal reabsorption**
absorption further down the intestinal tract of a substance or substances that have been absorbed before and subsequently excreted into the intestinal tract, usually through the bile.

- **intoxication**
1. poisoning, pathological process with clinical signs and symptoms caused by a substance of exogenous or endogenous origin. See **exogenous, endogenous.**
2. drunkenness following consumption of beverages containing ethanol or other compounds affecting the central nervous system.

- **inversion**
an atmospheric condition caused by increasing temperature with elevation, resulting in a layer of warm air preventing the rise of cooler air trapped beneath. This condition prevents the rise of pollutants that might otherwise be dispersed. Trapping pollutants near the ground increases ozone to harmful levels.

- **invert**
the lowest point of the channel inside a pipe, conduit, or canal.

- **ion**
an atom or radical in solution carrying an integral electric charge, either positive (cation) or negative (anion).

- **ion exchange**
a reversible chemical reaction between a solid (ion exchanger) and a fluid (usually a water solution) by means of which ions may be interchanged from one substance to another. The superficial physical structure of the solid is not affected.

- **ion exchange treatment**
a common water softening method often found on a large scale at water purification plants that remove some organics and radium by adding calcium oxide or calcium hydroxide to increase the pH to a level where the metals will precipitate out.

- **ionic concentration**
the concentration of any ion in solution, usually expressed in moles per litre.

- **ionic strength**
  a measure of the strength of a solution based on both the concentration and valences of the ions present.

- **ionisation**
  the splitting or dissociation (separation) of molecules into negatively and positively charged ions.

- **ionising radiation**
  any radiation consisting of directly or indirectly ionising particles or a mixture of both or photons with energy higher than the energy of photons of ultraviolet light or a mixture of both such particles and photons.

- **iron (Fe)**
  an essential element enabling the oxygen-carrying capacity of haemoglobin in red blood cells. A large portion of iron in the body is found in haemoglobin, muscle and other tissues, while some of the remaining iron is bound to an iron-transporting serum protein called transferrin. Tests to measure iron are performed to detect and evaluate iron-deficiency anaemias or abnormally high levels of iron found in haemochromatosis (excess iron deposits throughout the body).

- **irradiated food**
  food that has been briefly exposed to radioactivity (usually gamma rays) to kill insects, bacteria, and mould. Irradiated food can be stored without refrigeration or chemical preservatives and has a long 'shelf life'

- **irreversible alteration**
  change from normal structure or function that persists or progresses after cessation of exposure of the organism.

- **irreversible effect**
  effect characterised by the inability of the body to partially or fully repair injury caused by a toxic agent.

- **irritant**
  1. substance that causes inflammation following immediate, prolonged or repeated contact with skin, mucous membrane, or other biological material. A substance capable of causing inflammation on first contact is called a primary irritant.

  2. causing inflammation following immediate, pro-

longed or repeated contact with skin, mucous membrane or other tissues.

- **ischaemia**
local deficiency of blood supply and hence oxygen to an organ or tissue owing to constriction of the blood vessels or to obstruction.

- **isotonic**
denoting a fluid exerting the same osmotic pressure or water potential as another fluid with which it is being compared.

- **isotopic assay techniques**
methods of identifying and measuring substances such as hormones, by using procedures that employ labelling with material having radioactive properties (isotopes). Detection of radiation enables measurement of the desired analyte.

- **itai-itai disease**
illness observed in Japan possibly resulting from the ingestion of cadmium-contaminated rice. Damage occurred to the renal and skeleto-articular systems, the latter being very painful ('itai' means pain in Japanese).

- **jar test**
a laboratory procedure that simulates a water treatment plant's coagulation/flocculation units with differing chemical doses and also energy of rapid mix, energy of slow mix, and settling time. The purpose of this procedure is to estimate the minimum or ideal coagu- lant dose required to achieve certain water quality goals. Samples of water to be treated are commonly placed in six jars. Various amounts of chemicals are added to each jar, and the settling of solids is observed. The dose of chemicals that provides satisfactory settling removal of turbidity and/or colour is the dose used to treat the water being taken into the plant at that time. When evaluating the results of a jar test, the operator should also consider the floc quality in the flocculation area and the floc loading on the filter.

- **jaundice**
pathological condition characterised by deposition of bile pigment in the skin and mucous mem-

branes, including the conjunctivae, resulting in yellow appearance of the patient or animal. See **hyperbilirubinaemia, icterus**.

- **jet nebuliser**
  a nebuliser employing air pressure to aerosolise a bulk liquid.

- **jogging**
  the frequent starting and stopping of an electric motor.

- **joint effect**
  simultaneous or successive effect of factors of diverse types (chemical, physical, biological) on an organism. See **additive effect, antagonism, combined effect of poisons, potentiation, synergism**.

- **joule**
  a measure of energy, work or quantity of heat. One joule is the work done when a force of one Newton is displaced a distance of one meter in the direction of force.

- **karst**
  a geological formation of irregular limestone deposits with sinks, underground streams, and caverns.

- **ketoacidosis**
  a condition in which acidosis is accompanied by the accumulation of ketone bodies in the body tissues and fluids as in diabetic acidosis and starvation acidosis.

- **kg, kilogram**
  1,000 grams.

- **kilo**
  1. Kilogram. 2. Kilometer. 3. A prefix meaning 'thousand' used in the metric system and other scientific systems of measurement.

- **kinetic energy**
  energy possessed by a moving body of matter, such as water, as a result of its motion.

- **kinetic rate coefficient**
  a number that describes the rate at which a water constituent such as a biochemical oxygen demand or dissolved oxygen increases or decreases.

- **kinetics**
  the effects of forces on the motion of matter, of con-

cern with the velocity or speed of reactions.

- **kjeldahl nitrogen (kell-doll)**
  nitrogen in the form of organic proteins or their decomposition product ammonia, as measured by the Kjeldahl Method.

- **kleihauer-betke test**
  an indicator of foetal (F) haemoglobin in red cells. The test is based on the fact that haemoglobin F is not readily dissolved in an acid buffer, unlike normal adult haemoglobin (A). Demonstration of these two haemoglobins aids in the diagnosis of a number of blood disorders, and of foetal-maternal haemorrhage.

- **kraft**
  an alkaline chemical pulping process, using salt cake as makeup.

- **kw**
  kilowatt.

- **kw-h**
  kilowatt Hour.

- **lachrymator**
  see **lacrimator**

- **lacrimator**
  substance that irritates the eyes and causes the production of tears or increases the flow of tears.

- **lactate dehydrogenase or lactic acid dehydrogenase (LDH)**
  an enzyme important in provision of muscle energy (carbohydrate metabolism), that is present in many cells, particularly those of the heart, kidneys, liver, and skeletal muscle. Elevated blood levels of LDH may confirm a suspected heart attack.

- **lagoon**
  a shallow, artificial treatment pond where sunlight, bacterial action, and oxygen work to purify wastewater; a stabilisation pond. An aerated lagoon is a treatment pond that uses oxygen to speed up the natural process of biological decomposition of organic wastes.

- **LLN**
  lung associated lymph nodes.

- **laminar flow**
  gas flow with a smooth, non-turbulent pattern of streamlines, with no streamline

looping back on itself; usually occurs at very low Reynolds numbers.

- **landfill**
  a method for final disposal of solid waste on land. The refuse is spread and compacted and a cover of soil applied so that effects on the environment (including public health and safety) are minimised. Under current regulations, landfills are required to have liners and leachate treatment systems to prevent contamination of ground water and surface waters. An industrial landfill disposes of non-hazardous industrial wastes. A municipal landfill disposes of domestic waste including garbage, paper, etc. This waste may include toxins that are used in the home, such as insect sprays and powders, engine oil, paints, solvents, and weed killers.

- **langelier index (LI)**
  an index reflecting the equilibrium pH of a water with respect to calcium and alkalinity. This index is used in stabilsing water to control both corrosion and the deposition of scale. Langelier index = pH - pHs. where pH = actual pH of the water, and pHs= pH at which the water having the same alkalinity and calcium content is just saturated with calcium carbonate.

- **large water system**
  a water system that serves more than 50,000 persons

- **larvicide**
  substance intended to kill larvae.

- **laryngospasm**
  reflex spasmodic closure of the sphincter of the larynx, particularly the glottic sphincter.

- **larynx**
  main organ of voice production, the part of the respiratory tract between the pharynx and the trachea.

- **lassitude**
  weakness, exhaustion.

- **latency**
  time from the first exposure of a chemical until the appearance of a toxic effect.

- **latency period**
  the period of time between exposure to something that causes a disease and the onset of the health effect. Cancer caused by chemical exposure may have a latency period of 5 to 40 years.

- **latent effect**
  see **delayed effect**.

- **latent period**
  delay between exposure to a disease-causing agent and the appearance of manifestations of the disease: also defined as the period from disease initiation to disease detection. See **latency**.

- **laundering weir**
  sedimentation basin overflow weir. A plate with V-notches along the top to assure a uniform flow rate and avoid short-circuiting.

- **launders**
  sedimentation basin and filter discharge channels, consisting of overflow weir plates (in sedimentation basins) and conveying troughs.

- **lauric acid hydroxylase (hydroxylation)**
  lauric acid 11- (=omega-1) or 12 (=omega) hydroxylating enzyme which is used as a marker for the measurement of an isoenzyme of cytochrome P-450. This microsomal enzyme is induced in rodent liver by several peroxisomal proliferators. See **microsome, peroxisome**.

- **lavage**
  irrigation or washing out of a hollow organ or cavity such as the stomach, intestine or the lungs.

- **laxative**
  substance that causes evacuation of the intestinal contents. See **cathartic, purgative**.

- **LC_{50}**
  1. the statistically derived median lethal concentration, it is the concentration of a test material that can be expected to cause 50% mortality in a group of test animals for a given duration. Usually associated with an exposure duration, e.g., 4-hour LC_{50}.
  2. the concentration of a chemical in air or water which is expected to cause death in 50% of test animals living in that air or water.

• **LC_n**
this abbreviation is used for the exposure concentration of a toxicant lethal to n% of a test population. See median lethal concentration.

• **ldh**
see **lactate dehydrogenase**.

• **ldl cholesterol**
see **low density lipoprotein cholesterol.**

• **leachate**
liquid (mainly water) that percolates through a landfill and has picked up dissolved, suspended, and/or microbial contaminants from the waste. Leachate can be compared to coffee: water that has percolated down through the ground coffee.

• **leaching**
as water moves through soils or landfills, chemicals in the soil may dissolve in the water thereby contaminating the groundwater. This is called leaching.

• **leaching**
the process by which soluble substances are dissolved and transported down through the soil by recharge.

• **lead service line**
a service line made of lead which connects the water main to the building inlet and any lead pigtail, gooseneck or other fitting which is connected to such lead line.

• **lead(Pb)**
a heavy metal that is hazardous to health if breathed or swallowed. Its use in gasoline paints, and plumbing compounds has been sharply restricted or eliminated by central laws and regulations. See **heavy metals.**

• **legionella**
a genus of bacteria, some species of which have caused a type of pneumonia called Legionnaires Disease.

• **lesion**
1. area of pathologically altered tissue.
2. injury or wound.
3. infected patch of skin.

• **lethal**
deadly, fatal, causing death

- **lethal concentration 50 (LC_{50})**
  a concentration of a pollutant or effluent at which 50% of the test organisms die. A common measure of acute toxicity.

- **lethal dose 50 (LD_{50})**
  the dose of a toxicant that will kill 50% of test organisms within a designated period of time. The lower the LD_{50}, the more toxic the compound.

- **lethal synthesis**
  metabolic formation of a highly toxic compound from one that is relatively non-toxic (bioactivation), often leading to death of affected cells. See **suicide metabolism.**

- **leukaemia**
  progressive, malignant disease of the blood-forming organs, characterised by distorted proliferation and development of leucocytes and their precursors in the bone marrow and blood.

- **leukopenia**
  reduced concentration of leukocytes in the blood.

- **level controls**
  a float device (or pressure switch) which senses changes in a measured variable and opens or closes a switch in response to that change. In its simplest form, this control might be a floating ball connected mechanically to a switch or valve such as is used to stop water flow into a toilet when the tank is full.

- **lgK_{ow}**
  see **lgP_{ow}**.

- **lgP_{ow}**
  logarithm of base 10 of the partition coefficient of a substance between octan-1-ol and water. As an empirical measure for lipophilicity used for calculating bioaccumulation, fish toxicity, membrane adsorption and penetration etc. See **lipophilicity, octanol-water partition coefficient, partition coefficient.**

- **lh-rh**
  man-made compounds that are similar to the natural leutinising hormone-releasing hormone which aids in the production of testosterone.

- **life-cycle test**
  a comparative study in which organisms, that are subjected to different treatments, are observed at least from a life stage in one generation to the same life stage in the next generation.

- **life-long exposure**
  subjection to a potentially toxic substance during the whole lifetime.

- **ligase chain reaction (LCR)**
  a method of DNA amplification similar to PCR, except that it uses four primers instead of two and uses the enzyme ligase to ligate or join two segments of DNA. It can have greater specificity than PCR.

- **limacide**
  substance intended to kill mollusca including the gastropod mollusc, *Limax*.

- **limit recommended**
  see **recommended limit**.

- **limit test**
  acute toxicity test in which, if no ill-effects occur at a pre-selected maximum dose, no further testing at greater exposure levels is required. See **fixed dose test**.

- **limited degradation**
  a policy that allows for some lowering of natural environmental quality to a given level beneath an established health standard.

- **limited evidence**
  'limited evidence' is a collection of facts and accepted scientific inferences that suggests that an agent may be causing an effect, but this suggestion is not strong enough to be considered established fact. See carcinogenicity.

- **linearity**
  how closely an instrument measures actual values of a variable through its effective range. A measure used to determine the accuracy of an instrument.

- **linearised multi-stage model**
  sequence of steps in which (a) a multi-stage model is fitted to tumour incidence data, (b) the maximum linear term consistent with the data is calculated, (c) the low-dose slope of the dose-response function is equated to the coefficient of

the maximum linear term, and (d) the resulting slope is then equated to the upper bound of potency. See **multistage model.**

- **liner**
structure of natural clay or manufactured material (plastic) which serves as a barrier to restrict leachate from reaching or mixing with ground water in landfills, lagoons, etc.

- **lipase**
an enzyme originating in the pancreas which helps to digest fats. Measuring blood levels of lipase will help screen for disease of the pancreas.

- **lipophilic**
having an affinity for fat and high lipid solubility: a physicochemical property which describes a partitioning equilibrium of solute molecules between water and an immiscible organic solvent, favouring the latter, and which correlates with bioaccumulation. See **bioaccumulation, bioaccumulation factor, bioconcentration, octanol-water partition coefficient, hydrophobicity, hydrophilicity, lipophobicity**.

- **lipophobic**
having a low affinity for fat and a high affinity for water. See **bioaccumulation, bioaccumulation factor, bioconcentration, octanol-water partition coefficient, hydrophilicity, hydrophobicity, lipophilicity**.

- **liposome**
originally a lipid droplet in the endoplasmic reticulum of a fatty liver. Now applied to an artificially formed lipid droplet, small enough to form a relatively stable suspension in aqueous media and with potential use in drug delivery.

- **litter**
the highly visible portion of solid waste (usually packaging material) which is generated by the consumer and carelessly discarded outside of the regular garbage disposal system, as on the highways or in streets.

- **littoral zone**
1. That portion of a body of fresh water extending from the shoreline lakeward to

the limit of occupancy of rooted plants.
2. The strip of land along the shoreline between the high and low water levels.

- **liver nodule**
this medical term is applied to any small node or aggregation of cells within the liver.

- **loading**
the quantity of a substance entering the environment (soil, water, or air).

- **LOAEL**
lowest-observed-adverse-effect level, the lowest dose in an experiment which produced an observable adverse effect.

- **local effect**
circumscribed change occurring at the site of contact between an organism and a toxicant. See **systemic effect.**

- **LOEC (lowest-observed-effect-concentration)**
the lowest concentration of an effluent or toxicant that results in adverse effects on the test organisms (i.e., where the values for the observed endpoints are statistically different from the control).

- **logarithm**
the exponent that indicates the power to which a number must be raised to produce a given number. For example: if $B^2 = N$, the 2 is the logarithm of N (to the base B), or $10^2 = 100$ and $\log_{10} 100 = 2$. Also abbreviated to 'log'.

- **Logit transformation**
mathematical transformation that relates response to a stated dose or concentration of a toxicant to the response in the absence of the toxicant by the formula:
$\text{Logit} = \lg [B/(B_0-B)]$
where B is the response to the stated dose or concentration and B_0 is the response in the absence of the toxicant. Plotting the Logit function against the logarithm of base 10 of the dose or concentration usually gives a linear relationship.

- **lognormal size distribution**
particle size distribution characterised by a bell-shaped or Gaussian distribution shape when plotted on a logarithmic-size scale.

- **long-term averages (LTAs)**
  average of pollutant concentration or effluent toxicity calculated from waste load allocations (WLAs), typically assuming that the WLA is a 99th percentile value (or another upper bound value) based on the lognormal distribution. One LTA is calculated for each WLA (typically an acute LTA and a chronic LTA for aquatic life protection). The LTA represents the expected long-term average performance from the permitted facility required to achieve the associated WLA (EPA, 2002).

- **long-term effect**
  see **chronic effect**

- **long-term exposure**
  continuous or repeated exposure to a substance over a long period of time, usually of several years in man, and of the greater part of the total life-span in animals or plants. See **chronic exposure.**

- **low density lipoprotein cholesterol (LDLC)**
  the protein-lipid combination that transports the major amount of the cholesterol in the blood. While LDLC helps in synthesis of bile acid and steroid hormones, elevated blood levels of LDLC (the 'bad' cholesterol) have been linked to heart disease because it carries the cholesterol through the blood to the cells.

- **lower respiratory tract**
  those structures of the respiratory tract below the larynx.

- **lowest lethal concentration found**
  see **minimum lethal concentration.**

- **lowest-observed-effect-level (LOEL)**
  lowest concentration or amount of a substance, found by experiment or observation, that causes any alteration in morphology, functional capacity, growth, development, or life span of target organisms distinguishable from normal (control) organisms of the same species and strain under the same defined conditions of exposure. See **adverse effect, lowest-observed-adverse-effect-level, no-observed-ad-**

verse-effect-level, no-observed-effect-level.

- **lymph nodes**
small bean-shaped structures scattered along the vessels of the lymphatic system. The lymph nodes produce white blood cells and filter bacteria and cancer cells that may travel through the system.

- **lymphocyte**
animal cell that interacts with a foreign substance or organism, or one which it identifies as foreign, and initiates an immune response against the substance or organism. There are two groups of lymphocytes, B lymphocytes and T lymphocytes. See B lymphocyte, immune response, T lymphocyte.

- **lymphoma**
general term comprising tumours and conditions allied to tumours arising from some or all of the cells of lymphoid tissue.

- **lysimeter**
laboratory column cf selected representative soil or a protected monolith of undisturbed field soil with which it is possible to sample and monitor the movement of water and substances.

- **lysosome**
membrane-bound cytoplasmic organelle containing hydrolytic enzymes.

- **macrophage**
large (10-20 mm diameter) amoeboid and phagocytic cell found in many tissues, especially in areas of inflammation. Macrophages are derived from blood monocytes and play an important role in host defence mechanisms.

- **macroscopic (gross) pathology**
study in diseased tissues of changes which are visible to the naked eye.

- **macroscopic organisms**
organisms big enough to be seen by the eye without the aid of a microscope.

- **macrovascular disease - large vessel disease**
atherosclerosis involving the coronary, cerebrovascular, and peripheral vessels. Occurs at an earlier age and with greater frequency in those with diabetes than it

does in nondiabetic individuals.

- **mad hatter syndrome**
  see **mercurialism.**
- **magnesium (Mg)**
  an essential element which influences many enzymes needed to produce cellular energy and nerve and muscle message transmission. It affects nervous, muscular and cardiovascular systems. Magnesium is found mainly in bone, also in muscle and other tissues. Deficiency can lead to neuromuscular and central nervous system irritability, muscle twitches and weakness. Magnesium excess has been found to result in mental depression and cardiac arrest.
- **magnusson and kligman test**
  see **guinea-pig maximisation test.**
- **mainstream smoke (tobacco smoking)**
  smoke that is inhaled. See **sidestream smoke.**
- **malaise**
  vague feeling of bodily discomfort.
- **malignancy**
  population of cells showing both uncontrolled growth and a tendency to invade and destroy other tissues. A malignancy is life-threatening. See **cancer, metastasis, tumour**.
- **malignant tumour**
  a tumour which is cancerous.
- **mammogram**
  a low voltage X-ray photograph of the breast. It allows early detection of malignant or benign tumours and other microcalcification-containing lesions.
- **managerial controls**
  methods of non-point source pollution control that are derived from managerial decisions, such as changes in application times or rates for agrochemicals.
- **mania**
  emotional disorder (mental illness) characterised by an expansive and elated state (euphoria), rapid speech, flight of ideas, decreased need for sleep, distractability, grandiosity, poor judgement and increased motor activity.

- **manifold**
  a large pipe to which a series of smaller pipes are connected. Also called a Header.

- **manometer**
  an instrument for measuring pressure. Usually, a manometer is a glass tube filled with a liquid that is used to measure the difference in pressure across a flow-measuring device such as an orifice or Venturi meter. The instrument used to measure blood pressure is a type of manometer.

- **margin of exposure (MOE)**
  ratio of the no-observed-adverse-effect level (NOAEL) to the theoretical or estimated exposure dose (EED) or concentration (EEC). See **therapeutic index.**

- **margin of safety (MOS)**
  maximum amount of exposure producing no measurable effect in animals (or studied humans) divided by the actual amount of human exposure in a population.

- **mass (equivalent) diameter**
  diameter of a sphere composed of the particle bulk material with no voids that has the same mass as the particle in question.

- **mass mean diameter**
  diameter of a particle with a mass equal to the mean mass of all the particles in a population.

- **mass median aerodynamic diameter (MMAD)**
  the geometric mean aerodynamic diameter. Fifty per cent of the particles by weight will be smaller that the MMAD, 50% will be larger.

- **mass median diameter**
  diameter of a particle with the median mass of all the particles in a population.

- **mathematical model**
  model used during risk assessment to perform extrapolations.

- **maximum allowable (admissible, acceptable) concentration (MAC)**
  regulatory value defining the concentration that if inhaled daily (in the case of work people for 8 hours with a working week of 40 hours, in the case of the general population 24 hours) does not, in the

present state of knowledge, appear capable of causing appreciable harm, however long delayed during the working life or during subsequent life or in subsequent generations. See **permissible exposure limit, threshold limit value**.

- **maximum average daily concentration of an atmospheric pollutant**
highest of the average daily concentrations recorded at a definite point of measurement during a certain period of observation. See **peak daily average concentration of an air pollutant**.

- **maximum contaminant level (MCL)**
the highest (maximum) level of a contaminant allowed to go uncorrected by a public water system . Depending on the contaminant, allowable levels might be calculated as an average over time, or might be based on individual test results. Corrective steps are implemented if the MCL is exceeded.

- **maximum contaminant level goal (MCLG)**
the maximum level of a contaminant in drinking water at which no known or anticipated adverse effect on the health of persons would occur, and which allows an adequate margin of safety. Maximum contaminant level goals are non-enforceable health goals .

- **maximum permissible concentration (MPC)**
see **maximum allowable concentration**

- **maximum permissible daily dose**
maximum daily dose of substance whose penetration into a human body during a lifetime will not cause diseases or health hazards that can be detected by current investigation methods and will not adversely affect future generations.

- **maximum permissible level (MPL)**
level, usually a combination of time and concentration, beyond which any exposure of humans to a chemical or physical agent in their immediate environment is un-

safe. See **maximum allowable concentration**.

- **maximum residue limit (MRL)**
  MRLs are estimated for individual pesticides or veterinary drug residues in various food commodities. They are based on good agricultural practice (pesticides) or good practice in the use of veterinary drugs in which the product has been used in an efficacious manner and appropriate withdrawal periods have been followed. They are expressed as either the parent compound or a metabolite that is, or is representative of, the residue of toxicological concern in the food commodity. MRLs are not based upon toxicological data, but crude estimates of their toxicological significance are usually made by comparing the acceptable daily intake (ADI) with a calculation of the total intake of the residue based on the MRLs and food intake data of these commodities for which MRLs have been established.

- **maximum tolerable concentration (MTC)**
  highest concentration of a substance in an environmental medium that does not cause death of test organisms or species (denoted by LC_0).

- **maximum tolerable dose (MTD)**
  highest amount of a substance that, when introduced into the body, does not kill test animals (denoted by LD_0).

- **maximum tolerable exposure level (MTEL)**
  maximum amount or concentration of a substance to which an organism can be exposed without leading to an adverse effect after prolonged exposure time.

- **maximum tolerated dose (MTD)**
  high dose used in chronic toxicity testing that is expected on the basis of an adequate sub-chronic study to produce limited toxicity when administered for the duration of the test period. It should not induce (a) overt toxicity, for example appreciable death of cells or organ dysfunction, or (b) toxic manifestations that are

predicted materially to reduce the life span of the animals except as the result of neo-plastic development or (c) 10 % or greater retardation of body weight gain as compared with control animals. In some studies, toxicity that could interfere with a carcinogenic effect is specifically excluded from consideration.

- **maximum total trihalomethane potential (MTTP)**
  the maximum concentration of total trihalomethanes produced in a given water containing a disinfectant residual, after 7 days at 25 °C or above.

- **MBAS**
  methylene - blue - active substances. These substances are used in surfactants or detergents.

- **MCL**
  see **maximum contaminant level**.

- **mean life**
  average lifetime of a molecular, atomic, or nuclear system in a specified state. For an exponentially decaying system, it is the average time for the number of molecules, atoms or nuclei in a specified state to decrease by a factor of e, the base of natural logarithms. See **mean time, turnover time**.

- **mean time**
  see **mean life.**

- **measured variable**
  a characteristic or component part that is sensed and quantified (reduced to a reading of some kind) by a primary element or sensor.

- **mechanical joint**
  a flexible device that joins pipes or fittings together by the use of lugs and bolts.

- **media**
  elements of a surrounding environment that can be sampled for contamination; usually soil, water, or air. Plants, as well as humans (when sampling blood, urine, etc) and animals (such as sampling fish to update fish consumption advisories) can also be considered media. The singular of 'media' is 'medium'

- **median effective concentration (EC_{50})**
  statistically derived concentration of a substance in an

environmental medium expected to produce a certain effect in 50 % of test organisms in a given population under a defined set of conditions.

- **median effective dose (ED_{50})**
statistically derived dose of a chemical or physical agent (radiation) expected to produce a certain effect in 50 % of test organisms in a given population or to produce a half-maximal effect in a biological system under a defined set of conditions.

- **median lethal concentration (LC_{50})**
statistically derived concentration of a substance in an environmental medium expected to kill 50 % of organisms in a given population under a defined set of conditions.

- **median lethal dose (LD_{50})**
statistically derived dose of a chemical or physical agent (radiation) expected to kill 50 % of organisms in a given population under a defined set of conditions.

- **median lethal time (TL_{50})**
statistically derived average time interval during which 50 % of a given population may be expected to die following acute administration of a chemical or physical agent (radiation) at a given concentration under a defined set of conditions.

- **median narcotic concentration (NC_{50})**
statistically derived concentration of a substance in an environmental medium expected to cause narcotic conditions in 50 % of a given population under a defined set of conditions.

- **median narcotic dose (ND_{50})**
statistically derived dose of a substance expected to cause narcosis in 50 % of test animals under a defined set of conditions.

- **medical monitoring**
a set of medical tests and physical exams specifically designed to evaluate whether an individual's exposure could negatively affect that person's health.

- **medical waste**
all wastes from hospitals, clinics, or other health care facilities ('Red Bag Waste') that contain or have come into contact with diseased

tissues or infectious microorganisms. Also referred to as infectious waste which is hazardous waste with infectious characteristics, including contaminated animal waste, human blood and blood products, pathological waste, and discarded sharps (needles, scalpels, or broken medical instruments).

- **medium-size water system**
  a water system that serves greater than 3,300 and less than or equal to 50,000 person.

- **MEG**
  a procedure used for checking the insulation resistance on motors, feeders, buss bar systems, grounds, and branch circuit wiring. Also see **megger**.

- **megger**
  (from megohm). An instrument used for checking the insulation resistance on motors, feeders, buss bar systems, grounds, and branch circuit wiring. Also see **MEG**.

- **megohm**
  meg means one million, so 5 megohms means 5 million ohms. A megger reads in millions of ohms.

- **meiosis**
  process of 'reductive' cell division, occurring in the production of gametes, by means of which each daughter nucleus receives half the number of chromosomes characteristic of the somatic cells of the species. See **chromosome,diploid, gamete, haploid, miosis.**

- **meniscus**
  The curved top of a column of liquid (water, oil, mercury) in a small tube. When the liquid wets the sides of the container (as with water), the curve forms a valley. When the confining sides are not wetted (as with mercury), the curve forms a hill or upward bulge.

- **mercurialism**
  chronic poisoning caused by the excessive use of mercury, by breathing its vapour, or by exposure in mining or smelting processes. See **Mad Hatter syndrome**.

- **mesh**
  one of the openings or spaces in a screen or wo-

ven fabric. The value of the mesh is usually given as the number openings per inch. This value does not consider the diameter of the wire or fabric, therefore, the mesh number does not always have a definite relationship to the size of the hole.

- **mesocosm**
see **microcosm.**

- **mesothelioma**
malignant tumour of the mesothelium of the pleura, pericardium or peritoneum, that may be caused by exposure to asbestos fibres and some other fibres. See **tumour, malignant.**

- **mesotrophic**
reservoirs and lakes which contain moderate quantities of nutrients and are moderately productive in terms of aquatic animal and plant life.

- **metabolic activation**
bio-transformation of a substance of relatively low toxicity to a more toxic derivative. See **activation, bio-transformation, lethal synthesis, bioactivation**.

- **metabolic half-life (half-time)**
time required for one half of the quantity of a substance in the body to be metabolically transformed into a derivative or to be eliminated. See **metabolic half-time, clearance, elimination.**

- **metabolic model**
analysis and theoretical reconstruction of the way in which the body deals with a specific substance, showing the proportion of the intake that is absorbed, the proportion that is stored and in what tissues, the rate of breakdown in the body and the subsequent fate of the metabolic products, and the rate at which it is eliminated by different organs as unchanged substance or metabolites.

- **metabolic transformation**
biochemical transformation of a substance that takes place within an organism. See **bio-transformation.**

- **metabolism**
1. sum total of all physical and chemical processes that take place within an organism. In a narrower sense, the physical and

chemical changes that take place in a given substance within an organism. It includes the uptake and distribution within the body of chemical compounds, the changes (biotransformation) undergone by such substances, and the elimination of the compounds and of their metabolites. See **biotransformation.**

2. the conversion or breakdown of a substance from one form to another by a living organism.

3. the sum of the chemical reactions occurring within a cell or a whole organism; includes the energy-releasing breakdown of molecules (catabolism) and the synthesis of new molecules (anabolism).

- **metabolite**

any intermediate or product resulting from metabolism.

- **metalimnion**

the middle layer in a thermally stratified lake or reservoir. In this layer there is a rapid decrease in temperature with depth. Also called the thermocline.

- **metaplasia**

abnormal transformation of an adult, fully differentiated tissue of one kind into a differentiated tissue of another kind. See **hyperplasia, neoplasia**.

- **metastasis**

1. movement of bacteria or body cells, especially cancer cells, from one part of the body to another, resulting in change in location of a disease or of its symptoms from one part of the body to another.

2. growth of pathogenic micro-organisms or of abnormal cells distant from the site of their origin in the body.

- **methaemoglobinaemia**

1.presenceof methaemoglobin (oxidised haemoglobin) in the blood in greater than normal proportion.

2. the presence of methaemoglobin in the bloodstream caused by the reaction of materials with the haemoglobin in red blood cells that reduces their oxygen-carrying capacity. Methaemoglobin is a soluble, brown, crystal-

line blood pigment that differs from haemoglobin in that it contains ferric iron and is unable to combine reversibly with molecular oxygen.

- **methaemoglobin-forming substance**
substance capable of oxidising directly or indirectly the iron(II) in haemoglobin to iron(III) to form methaemoglobin, a derivative of haemoglobin that cannot transport oxygen.

- **methane (CH_4)**
a colourless, non-poisonous, inflammable gas emitted by marshes and dumps undergoing anaerobic decomposition.

- **methoxychlor (meth-oxy-klor)**
a pesticide which causes adverse health effects in domestic water supplies and is also toxic to freshwater and marine aquatic life. The chemical name for methoxychlor is 2,2-bis (P-methoxyphenol)- 1, 1, I - trichloroethane.

- **methyl orange alkalinity**
a measure of the total alkalinity in a water sample. The alkalinity is measured by the amount of standard sulfuric acid required to lower the pH of the water to a pH level of 4.5, as indicated by the change in color of methyl orange from orange to pink. Methyl orange alkalinity is expressed as milligrams per litre equivalent calcium carbonate.

- **mg**
milligram (1/1000, 10^{-3}, of a gram).

- **mg/cm2**
milligram per square centimeter (of a surface).

- **mg/kg**
milligrams per kilogram. Dosage used in toxicology testing to indicate a dose administered per kg of body weight (50 mg = 1 teaspoonful).

- **mg/l**
see **milligrams per litre**.

- **mg/mv**
milligram per cubic meter; a measure of the concentration of a chemical in a known volume (a cubic meter) of air, soil, or water.

- **microalbuminuria**
chronic presence of albumin in slight excess in urine.

- **microbial growth**
  the activity and growth of micro-organisms such as bacteria, algae, diatoms, plankton and fungi.

- **microcosm**
  artificial test system that simulates major characteristics of the natural environment for the purposes of ecotoxicological assessment. Such a system would commonly have a terrestrial phase, with substrate, plants and herbivores, and an aquatic phase, with vertebrates, invertebrates and plankton. The term 'mesocosm' implies a more complex and larger system than the term 'microcosm' but the distinction is not clearly defined. See **experimental model ecosystem.**

- **microgram (μg)**
  one-millionth of a gram

- **micrograms per litre (μg/l)**
  one microgram of a substance dissolved in each litre of water. This unit is equal to parts per billion (ppb) since one litre of water is equal in weight to one billion micrograms.

- **micromercurialism**
  effects of exposure to mercury detected at the lowest exposure levels producing a measurable reaction. See **mercurialism.**

- **micron**
  a unit of length. One millionth of a meter or one thousandth of a millimeter. One micron equals 0.00004 of an inch.

- **micro-organisms**
  bacteria, yeast, simple fungi, algae, protozoan, and a number of other organisms that are microscopic in size. Most are beneficial but some produce disease. Others are involved in composting and sewage treatment.

- **microsome**
  artefactual spherical particle, not present in the living cell, derived from pieces of the endoplasmic reticulum present in homogenates of tissues or cells: microsomes sediment from such homogenates when centrifuged at *100 000 g* and higher: the microsomal fraction obtained in this way is often used as a source of monooxygenase enzymes. See **cy-**

tochrome P-420, cytochrome P-448, cytochrome P-450, endoplasmic reticulum, mono-oxygenase, phase 1 reactions.

- **microvascular disease - small vessel disease**
  a group of disorders related to disease of small blood vessels particularly retinopathy and nephropathy characteristic of diabetes mellitus.

- **micturitic**
  see **diuretic.**

- **migration**
  moving from one location to another.

- **mil**
  a unit of length equal to 0.001 of an inch. The diameter of wires and tubing is measured in mils, as is the thickness of plastic sheeting

- **milligram (mg)**
  one-thousandth of a gram (3.5×10^{-1} oz. 0.000035 oz.).

- **milligrams per litre (mg/l)**
  a measure of concentration of a dissolved substance. A concentration of one mg/l means that one milligram of a substance is dissolved in each litre of water. For practical purposes, this unit is equal to parts per million (ppm) since one litre of water is equal in weight to one million milligrams. Thus a litre of water containing 10 milligrams of calcium has 10 parts of calcium per one million parts of water, or 10 parts per million (10 ppm).

- **millimicron**
  a unit of length equal to 10^{-3} microns (one thousandth of a micron), 10^{-6} mm or 10^{-9} meters. Correctly called a nanometer, nm.

- **million-gallons per day (mgd_{-1})**
  a measure of water flow.

- **minamata disease**
  neurological disease caused by ingestion of methyl mercury-contaminated fish, first seen at Minamata Bay in Japan.

- **mineralisation**
  the microbial conversion of an element from an organic to an inorganic state.

- **minimal risk level (MRL)**
  an estimate of daily human exposure to a hazardous substance at or below which that substance is unlikely to pose a measur-

able risk of harmful (adverse), non-cancerous effects. MRLs are calculated for a route of exposure (inhalation or oral) over a specified time period (acute, intermediate, or chronic). MRLs should not be used as predictors of harmful (adverse) health effects [see **reference dose**].

- **minimisation**
  measures or techniques that reduce the amount of wastes generated during industrial production processes. This term is also applied to recycling and other efforts to reduce the volume of waste going to landfills. This term is interchangeable with waste reduction and waste minimisation.

- **minimum lethal concentration (LC_{min})**
  lowest concentration of a toxic substance in an environmental medium that kills individual organisms or test species under a defined set of conditions. See **lowest lethal concentration found**.

- **minimum lethal dose (LD_{min})**
  lowest amount of a substance that, when introduced into the body, may cause death to individual species of test animals under a defined set of conditions.

- **minimum significant difference (MSD)**
  the minimum difference which can exist between a test treatment and the controls in a particular test and be statistically significant; a measure of test sensitivity. The lower the MSD the more sensitive the test. Related to test power, the MSD is decreased by increasing the number of replicates and/or decreasing the amount of between-replicate variability in the controls and treatments.

- **minute volume**
  the average volume of air (or gas) inspired or exhaled per minute. For physiological evaluations, expired minute volume is usually expressed as BTPS conditions. For calculating estimated doses, inspired minute volumes are usually

expressed as ATPD. Expired minute volumes are slightly less than inspired minute volumes.

- **meiosis**
  abnormal contraction of the pupil of the eye to less than 2 mm.

- **miscible**
  liquid substances capable of mixing without separation into two phases. Refers to liquid mixtures.

- **mist**
  a liquid particle aerosol, typically formed by physical shearing of liquids, such as in nebulisation, spraying, or bubbling.

- **mitigation**
  measures taken to reduce adverse effects on the environment.

- **mitochondrion**
  eukaryote cytoplasmic organelle that is bounded by an outer membrane and an inner membrane. The inner membrane has folds called cristae that are the centre of ATP synthesis in oxidative phosphorylation in the animal cell and supplement ATP synthesis by the chloroplasts in photosynthetic cells. The mitochondrial matrix within the inner membrane contains ribosomes, many oxidative enzymes, and a circular DNA molecule that carries the genetic information for a number of these enzymes.

- **mitogen**
  substance that induces lymphocyte transformation or, more generally, mitosis and cell proliferation. See **transformation**.

- **mitosis**
  process by which a cell nucleus divides into two daughter nuclei, each having the same genetic complement as the parent cell. Nuclear division is usually followed by cell division.

- **mixed function oxidases**
  oxidizing enzymes which are involved in the metabolism of many foreign compounds, giving products of different toxicity from the parent compound.

- **mixed liquor**
  activated sludge and water containing organic matter being treated in an aeration tank.

- **mixing zone**
  area where an effluent discharge undergoes initial dilution and is extended to cover the secondary mixing in the ambient water body. A mixing zone is an allocated impact zone where water quality criteria can be exceeded as long as acutely toxic conditions are prevented .

- **mixture**
  a heterogeneous association of materials that cannot be represented by a chemical formula and that does not undergo chemical change as a result of interaction amongst the mixed materials. The constituent materials may or may not be uniformly dispersed and can usually be separated by mechanical means (as opposed to a chemical reaction). Uniform liquid mixtures are called solutions

- **mobility (equivalent) diameter**
  diameter of a spherical particle with the same dynamic mobility as the particle in question.

- **modelling**
  1. use of mathematical equations to simulate and predict real events and processes.
  2. an investigative technique using a mathematical or physical representation of a system or theory that accounts for all or some of its known properties. Models are often used to test the effect of changes of system components on the overall performance of the system.

- **modifying factor (MF)**
  uncertainty factor that is greater than zero and less than, or equal to 10. The magnitude of the factor depends upon the professional assessment of scientific uncertainties of a study or database not explicitly treated with the standard uncertainty factors (for example the completeness of the overall database and the number of animals tested). The default value for the factor is 1. See **uncertainty factor**.

- **moisture content**
the water loss of a fully hydrated resin under controlled drying conditions.

- **molar or molarity**
a molar solution consists of one gram molecular weight of a compound dissolved in enough water to make one litre of solution. A gram molecular weight is the molecular weight of a compound in grams. For example, the molecular weight of sulfuric acid (H_2SO_4) is 98. A one M solution of sulfuric acid would consist of 98 grams of H_2SO_4 dissolved in enough distilled water to make one litre of solution.

- **mole**
a unit weight or volume of a chemical corresponding to its molecular weight. A mole of water weighs 18 g, and its vapour occupies 22.4 l at standard temperature and pressure.

- **molecular formula**
a written representation, using symbols, of a chemical entity. It shows the actual number and kind of atoms in a molecule.

- **molecular weight**
1. the mass in grams per mole of a substance. See **mole**.
2. the molecular weight of a compound in grams is the sum of the atomic weights of the elements in the compound. The molecular weight of sulfuric acid (H_2SO_4) in grams is 98.

- **molecule**
the smallest particle of an element or compound retaining its characteristics.

- **molluscicide**
substance intended to kill molluscs. See **limacide**.

- **monitoring**
1. continuous or repeated observation, measurement, and evaluation of health and/or environmental or technical data for defined purposes, according to prearranged schedules in space and time, using comparable methods for sensing and data collection. Evaluation requires comparison with appropriate reference values based on knowledge of the probable relationship between ambient exposure and adverse effects. See

ambient monitoring, biological effect monitoring, biological monitoring, environmental monitoring, health surveillance, personal monitoring.
2. periodic or continuous sampling to determine the level of pollution or radioactivity.

- **monitoring well**
a well used to take water quality samples or to measure ground water levels.

- **monoclonal**
pertaining to a specific protein from a single clone of cells, all molecules of this protein being the same.

- **monoclonal antibodies**
specific antibodies derived from a single (mono) clone of plasma cells. It may be produced in large quantities, usually for use against a specific antigen, for diagnostic and possibly therapeutic use.

- **monoclonal antibody**
antibody produced by cloned cells derived from a single lymphocyte. See **polyclonal antibody**.

- **monodisperse distribution**
a distribution of particles with a single size or a small range of sizes.

- **monomer**
a molecule of low molecular weight capable of reacting with identical or different monomers to form Polymers.

- **monomictic**
lakes and reservoirs which are relatively deep, do not freeze over during the winter months, and undergo a single stratification and mixing cycle during the year. These lakes and reservoirs usually become destratified during the mixing cycle, usually in the fall of the year.

- **mononuclear phagocyte system**
see **reticuloendothelial system.**

- **mono-oxygenase**
enzyme that catalyses reactions between an organic compound and molecular oxygen in which one atom of the oxygen molecule is incorporated into the organic compound and one atom is reduced to water. It is involved in the metabo-

lism of many natural and foreign compounds giving both non-reactive products and products of different or increased toxicity from that of the parent compound: such enzymes are the main catalysts of phase 1 reactions in the metabolism of xenobiotics by the endoplasmic reticulum or by preparations of microsomes. See **mixed function oxidase, cytochrome P-420, cytochrome P-448, cytochrome P-450, endoplasmic reticulum, microsome, phase 1 reactions.**

- **monovalent**
  having a valence of one, such as the cuprous (copper) ion, Cu^+.

- **morbidity**
  any departure, subjective or objective, from a state of physiological or psychological well-being: in this sense, 'sickness', 'illness' and 'morbid condition' are similarly defined and synonymous.

- **morbidity rate**
  term used loosely to refer to incidence or prevalence rates of disease.

- **morbidity survey**
  method for the estimation of the prevalence and/or incidence of a disease or diseases in a population. A morbidity survey is usually designed simply to ascertain the facts as to disease distribution, and not to test a hypothesis.

- **mordant**
  substance that fixes a dyestuff in or on a material by combining with the dye to form an insoluble compound, used to fix or intensify stains in a tissue or cell preparation.

- **mortality**
  death as studied in a given population or subpopulation. The word mortality is often used incorrectly instead of mortality rate.

- **mortality rate**
  see **death rate.**

- **mortality study**
  investigation dealing with death rates or proportion of deaths attributed to specific causes as a measure of response.

- **MOS**
  see **margin of safety.**

- **most probable number (MPN)**
MPN is the Most Probable Number of coliform-group organisms per unit volume of sample water. Expressed as the number of organisms per 100 ml of sample water.

- **motile**
capable of self-propelled movement. A term that is sometimes used to distinguish between certain types of organisms found in water.

- **motor efficiency**
the ratio of energy delivered by a motor to the energy supplied to it during a fixed period or cycle. Motor efficiency ratings will vary depending upon motor manufacturer and usually will range from 88.9 to 90.0 percent.

- **MPN**
see **most probable number**.

- **MTD**
maximum tolerated dose, the dose that an animal species can tolerate for a major portion of its lifetime without significant impairment or toxic effect other than carcinogenicity.

- **mudballs**
material that is approximately round in shape and varies from pea-sized up to two or more inches in diameter. This material forms in filters and gradually increases in size when not removed by the backwashing process.

- **mulch**
any substance spread or allowed to remain on the soil surface to conserve soil moisture and shield soil particles from the erosive forces of raindrops and runoff.

- **multigeneration study**
1. toxicity test in which two to three generations of the test organism are exposed to the substance being assessed.
2. toxicity test in which only one generation is exposed and effects on subsequent generations are assessed.

- **multiple (or multiphasic) screening**
procedure that has evolved by combining single screening tests, and is the logical corollary of mass screening. Where much time and effort have been spent by a popu-

lation in attending for a single test such as mass radiography, it is natural to consider the economy of offering other tests at the same time. Multiple (or multiphasic) screening implies the administration of a number of tests, in combination, to large groups of people. See **screening**.

- **multiple use**
  use of land for more than one purpose; i.e., grazing of livestock, wildlife production, recreation, watershed, and timber production. Could also apply to use of bodies of water for recreational purposes, fishing, and water supply.

- **multi-stage cluster sampling**
  cluster sampling with more than two stages, each sampling being made on aggregates (or clusters) in which the clusters already obtained by the preceding sampling have been divided.

- **multi-stage model**
  1. dose-response model for cancer death estimation of the form
  $P(d) = 1 - \exp[-(q_o + q_1 d^1 + q_2 d^2 + \ldots\ldots\ldots + q(k)d^k)]$
  where *P(d)* is the probability of cancer death from a continuous dose rate, *d*, the *q*'s are constants, and *k* is the number of dose groups (or, if less than the number of dose groups, *k* is the number of biological stages believed to be required in the carcinogenesis process). With the multi-stage model, it is assumed that cancer is initiated by cell mutations in a finite series of steps. A one-stage model is equivalent to a one-hit model.
  2. mathematical model based on the multi-stage theory of the carcinogenic process, which yields risk estimates either equal to or less than the one-hit model.

- **multi-stage pump**
  a pump that has more than one impeller A single-stage pump has one impeller.

- **multi-stage sampling**
  type of sampling in which the sample is selected by stages, the sampling units at each stage being subsampled from the larger units chosen at the previous stage.

- **municipal sewage**
wastes (mostly liquid) originating from a community. It may be composed of domestic wastewaters and/or industrial wastewaters.

- **murine**
of or belonging to the family of rats and mice (*Muridae*).

- **mutagen**
any substance that can induce heritable changes (mutations) of the genotype in a cell as a consequence of alterations or loss of genes or chromosomes (or parts thereof).

- **mutagenesis**
1. introduction of heritable changes (mutations) of the genotype in a cell as a consequence of alterations or loss of genes or chromosomes (or parts thereof).
2. production of mutations. This may lead to transformation and carcinogenesis. See **carcinogenesis**, **transformation**.

- **mutagenic**
causing mutations. Mutagenic substances may also be carcinogenic.

- **mutagenicity**
ability of a physical, chemical, or biological agent to induce heritable changes (mutations) in the genotype in a cell as a consequence of alterations or loss of genes or chromosomes (or parts thereof).

- **mutation**
any relatively stable heritable change in genetic material that may be a chemical transformation of an individual gene (gene or point mutation), altering its function, or a rearrangement, gain or loss of part of a chromosome, that may be microscopically visible (chromosomal mutation). Mutation can be either germinal and inherited by subsequent generations, or somatic and passed through cell lineage by cell division. See **chromosome**, **gene**, **clastogenesis**, **genotoxicity**.

- **myasthenia**
muscular weakness.

- **mycotoxin**
toxin produced by a fungus.

- **mydriasis**
extreme dilation of the pupil of the eye, either as a

result of normal physiological response or in response to a chemical exposure.

• **myelosuppression**
reduction of bone marrow activity leading to a lower concentration of platelets, red cells and white cells in the blood.

• **n**
see **normal**.

• **narcotic**
1. nonspecific usage - an agent that produces insensibility or stupor.
2. specific usage - an opioid, any natural or synthetic drug that has morphine-like actions.

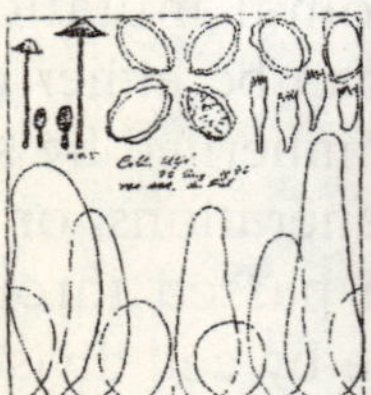

• **nasopharyngeal compartment**
a portion of the respiratory tract between the epiglottis and the anterior nares.

• **natriuretic**
substance increasing the rate of excretion of sodium ion in the urine.

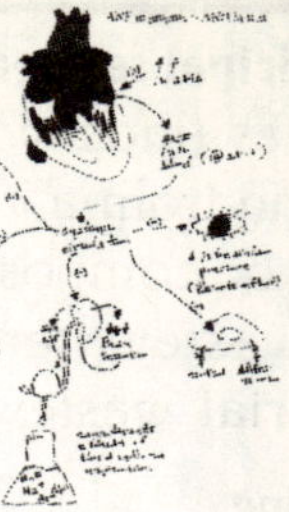

• **natural occurrence**
presence of a substance in nature, as distinct from presence resulting from inputs from human activities. The contamination of the natural environment by some man-made compounds may be so widespread that it is practically impossible to get access to biota with a truly natural level, only 'normal' levels can be measured, those which are usually prevalent in places where there is no obvious local contamination.

• **NCWS**
see **non-community water system**.

• **near the first service connection**
at one of the 20 percent of all service connections in the entire system that are nearest the water supply treatment facility, as mea-

sured by water transport time within the distribution system.

- **nebuliser**
  a device in which droplet aerosols are produced by dispersion of a bulk liquid.

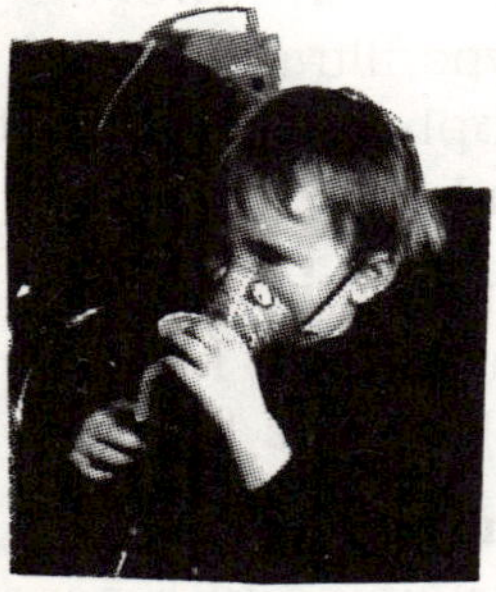

- **necropsy**
  see **autopsy**, **biopsy**.

- **necrosis**
  mass death of areas of tissue or bone surrounded by healthy areas.
  Morphological changes that follow cell death, characterised most frequently by nuclear changes.

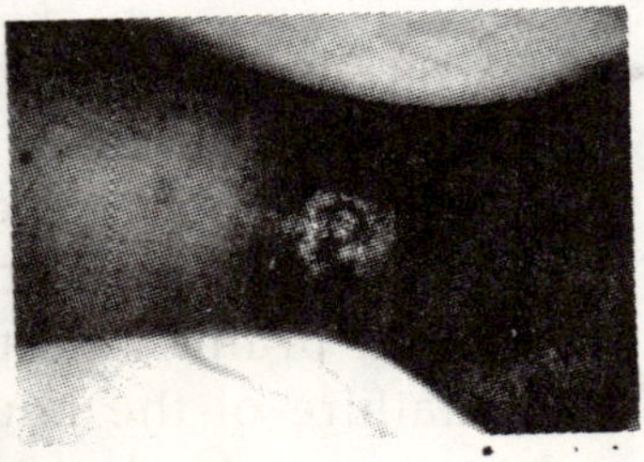

- **negligible risk**
  1. probability of adverse effects occurring that can reasonably be described as trivial.
  2. probability of adverse effects occurring that is so low that it cannot be reduced appreciably by increased regulation or investment of resources.
  See **acceptable risk**, **accepted risk**, **risk *de minimis***.

- **nematicide**
  chemical used to kill nematodes.

- **nematodes**
  roundworms, any of which are pathogenic for plants and sometimes animals.

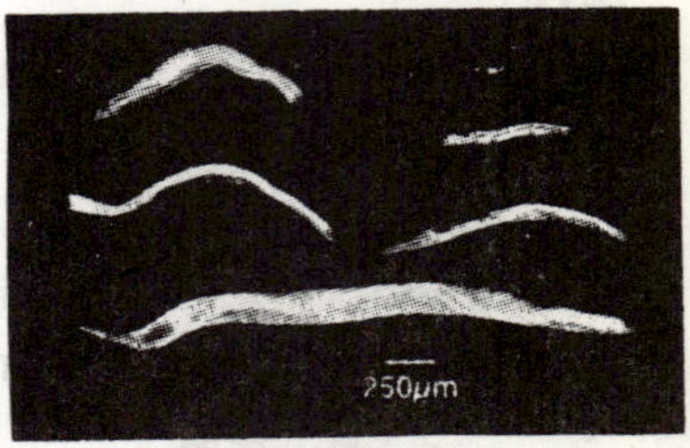

- **neonate**
  infant during the first 4 weeks of postnatal life. For statistical purposes some scientists have defined the period as the first 7 days.

• **neoplasia**
new and abnormal formation of tissue as a tumour or growth by cell proliferation that is faster than normal and continues after the initial stimulus (i) that initiated the proliferation has ceased. See **tumour, hyperplasia, metaplasia**.

• **nephelometric**
a means of measuring turbidity in a sample by using an instrument called a nephelometer. A nephelometer passes light through a sample and the amount of light deflected (usually at a 90- degree angle) is then measured.

• **nephelometric turbidity unit (NTU)**
the unit of measure for turbidity.

• **nephritis**
inflammation of the kidney, leading to kidney failure,

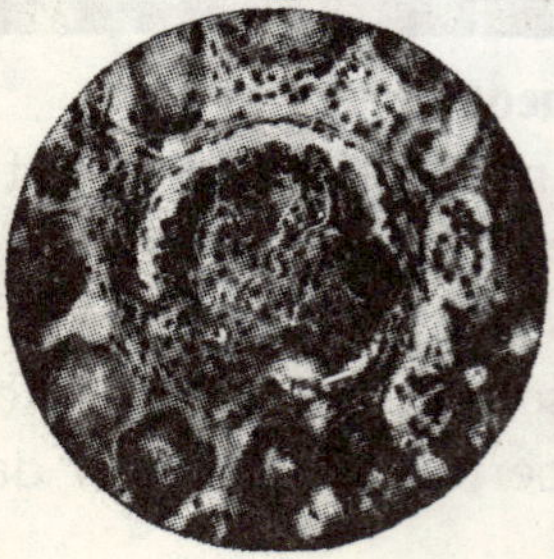

usually accompanied b proteinuria, haematuria oedema, and hypertension

• **nephropathy**
a disease of the kidneys Commonly accompanie later stages of diabetes me litus. It begins wit hyperfiltration, renal hyper trophy, microalbuminuria and hypertension.

• **nephrotoxic**
chemically harmful to th cells of the kidney.

• **neural**
pertaining to a nerve or t the nerves.

• **neural tube**
the tubular embryonic tis sue from which the brai and spinal cord develop Very serious congenital de fects result from failure o this tube to close during the process of growth and dif ferentiation.

• **neural tube defect**
a congenital defect of the central nervous system including the spinal cord skull and brain, resulting from failure of the neura tube to properly close dur ing foetal development

Defects may include absence of the skull, and protrusions of the brain or spinal cord. Most such defects can be detected before birth by determination of amniotic fluid or blood levels of alpha-foetoprotein and by ultrasonic scanning.

- **neuron(e)**

nerve cell, the morphological and functional unit of the central and peripheral nervous systems.

- **neuropathy**

any disease of the central or peripheral nervous system.

- **neurotoxic**

able to produce chemically an adverse effect on the nervous system. Such effects may be subdivided into two types.

Central nervous system effects (including transient effects on mood or performance and pre-senile dementia such as Alzheimer's disease).

Peripheral nervous system effects (such as the inhibitory effects of organophosphorus compounds on synaptic transmission).

- **neutralisation**

the chemical process in which the acidic or basic characteristics of a fluid are changed to those of water (pH = 7).

- **newton**

a force which, when applied to a body having a mass of one kilogram, gives it an acceleration of one metre per second per second.

- **nitrate**

a compound containing nitrogen which can exist in the atmosphere or as a dissolved gas in water and which can have harmful effects on humans and animals. Nitrates in water can cause severe illness in infants and cows.

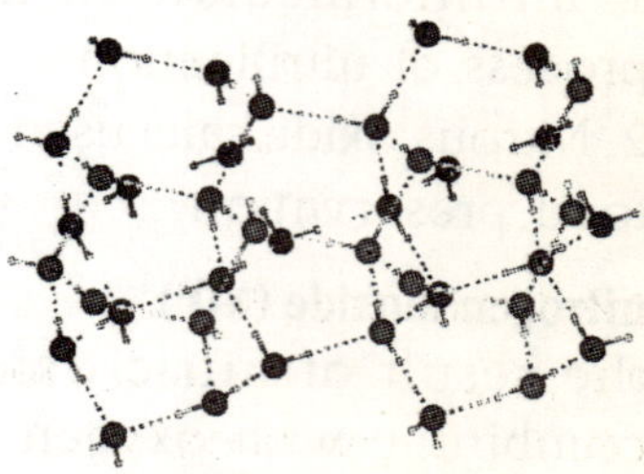

- **nitric oxide (NO)**

a gas formed by combustion under high temperature and high pressure in an internal combustion engine. It changes into nitrogen dioxide in the ambient air and contributes to photochemical smog.

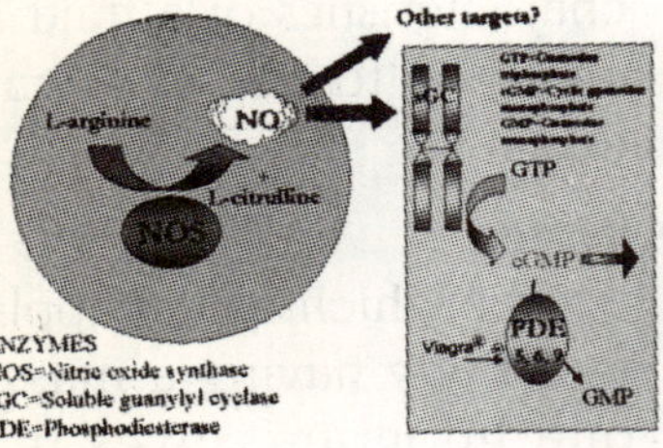

- **nitrification**
  sequential oxidation of ammonium salts to nitrite and nitrate by micro-organisms.

- **nitrification inhibitor**
  a chemical that slows down the conversion of ammonium to nitrate nitrogen.

- **nitrite**
  1. an intermediate in the process of nitrification.
  2. Nitrous oxide salts used in food preservation.

- **nitrogen dioxide (NO²)**
  the result of nitric oxide combining with oxygen in the atmosphere, a major component of photochemical smog.

- **nitrogen fixation**
  the biological or chemical process by which elemental nitrogen, from the air, is converted to organic or available nitrogen.

- **nitrogen oxide (NO²)**
  product of combustion from transportation and stationary sources and a major contributor to the formation of ozone in the troposphere and acid deposition.

- **nitrogenous**
  a term used to describe chemical compounds (usually organic) containing nitrogen in combined forms. Proteins and nitrates are nitrogenous compounds.

- **nitrogenous wastes**
  animal or plant residues that contain large amounts of nitrogen.

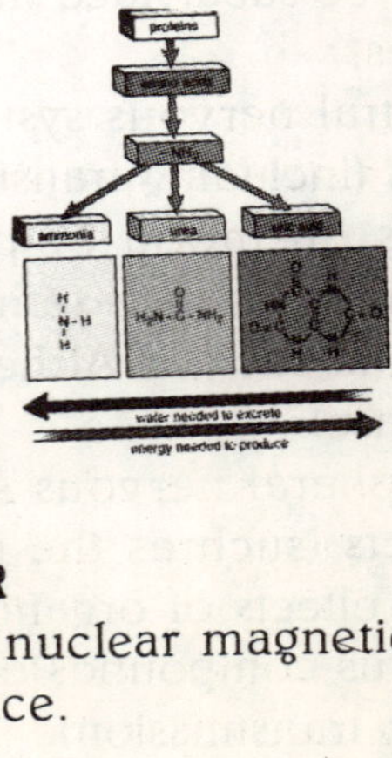

- **NMR**
  see nuclear magnetic resonance.

- **no acceptable daily intake allocated**
  this expression is applicable to a substance for which the available information is not sufficient to establish its safety, or when the specifications for identity and purity are not adequate, or when the available data show that the substance is hazardous and should not be used. The basis for the use of the expression should be determined before action is taken. In the first two cases above, not being able to allocate an ADI does not mean that the substance is unsafe. See **acceptable daily intake**.

- **no apparent public health hazard**
  a category used for sites where human exposure to contaminated media might be occurring, might have occurred in the past, or might occur in the future, but where the exposure is not expected to cause any harmful health effects.

- **no effect level (NEL)**
  maximum dose (of a substance) that produces no detectable changes under defined conditions of exposure. At present, this term tends to be substituted by no-observed-adverse-effect-level (NOAEL) or no-observed-effect-level (NOEL).

- **no observed effect level (NOEL)**
  this is the greatest concentration or amount of a chemical found by experiment or observation that causes no detectable adverse alteration of morphology, functional capacity, growth, development, or life span of the target organism. The maximum dose or ambient concentration which an organism can tolerate over a specified period of time without showing any detectable adverse effect and above which adverse effects are apparent. See **adverse effect**.

- **no public health hazard**
  a category used for sites where people have never and will never come into contact with harmful amounts of site-related substances.

- **NOAEC (not-observed-acute-effect-concentration)**
  the NOEC for acute effects (typically survival); see **NOEC**.

- **NOAEL**
  no-observed-adverse-effect level. The highest dose in an experiment which did not produce an observable adverse effect.

- **noble metal**
  chemically inactive metal (such as gold). A metal that does not corrode easily and is much scarcer (and more valuable) than the so-called useful or base metals. Also see **base metal**.

- **n-octanol-water partition coefficient**
  see **octanol-water partition coefficient**.

- **nodule**
  small node or boss that is solid and can be detected by touch.

- **NOEC (no-observed-effect-concentration)**
  the highest concentration of an effluent or toxicant that causes no observable adverse effects on the test organisms (i.e., the highest concentration of toxicant at which the values for the observed responses are not statistically different from the control)

- **nominal concentration**
  a gross estimate of exposure concentration based upon test material usage. The total amount of test material used during a given exposure day divided by the total volume of air passed through the chamber during the exposure.

- **nominal diameter**
  an approximate measurement of the diameter of a pipe. Although the nominal diameter is used to describe the size or diameter of a pipe, it is usually not the exact inside diameter of the pipe.

- **non-bioenvironmental transformation**
  change in the chemical or physical nature of a substance occurring as a result of physiochemical conditions and independent of any biological system.

- **non-community water system (NCWS)**
  a public water system that is not a community water system. There are two types of NCWSs: transient and non-transient.

- **non-contact cooling water**
  water used for cooling which does not come into direct contact with any raw material, product, by-product, or waste.

- **non-conventional pollutant**
  any pollutant which is not statutorily listed or which is poorly understood by the scientific community.

- **non-effective dose**
  amount of a substance that has no effect on the organism. It is lower than the threshold of harmful effect and is estimated while establishing the threshold of harmful effect. See **sub-threshold dose, threshold**.

- **non-ionic polymer**
  a polymer that has no net electrical charge.

- **non-isotopic immunoassay**
  a method of identifying and measuring substances, such as hormones, by utilising antigen-antibody reactions without the use of radioactive material (isotopes) for labelling. Such assays may employ other kinds of reaction-monitoring signals.

- **non-occupational exposure**
  environmental exposure outside the workplace to substances that are otherwise associated with particular work environments and/or activities and processes that occur there.

- **non-parametric hypothesis tests**
  hypothesis test methods that are not dependent on data being normally distributed and exhibiting homogeneity of variance. These are also called distribution-free methods.

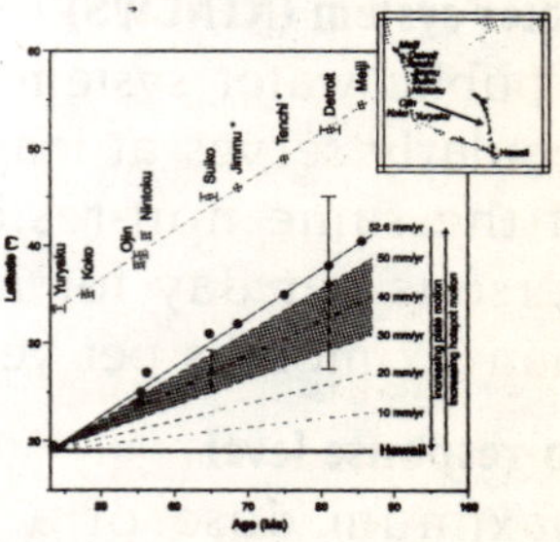

- **non-point source**
  any source of pollution not associated with a distinct discharge point. Includes

sources such as rainwater, runoff from agricultural lands, industrial sites, parking lots, and timber operations, as well as escaping gases from pipes and fittings.

- **non-potable**
  water that may contain objectionable pollution, contamination, minerals, or infective agents and is considered unsafe and/or unpalatable for drinking.

- **non-target organism**
  organism affected by a pesticide although not the intended object of its use.

- **non-thyroidal illness (NTI)**
  illness that causes abnormal thyroid function test results in the absence of thyroid disease.

- **non-transient non-community water system (NTNCWS)**
  a public water system that regularly serves at least 25 of the same non-resident persons per day for more than six months per year.

- **no-response level**
  maximum dose of a substance at which no specified response is observed in a defined population and under defined conditions of exposure.

- **normal**
  a normal solution contains one gram equivalent weight of reactant (compound) per litre of solution. The equivalent weight of an acid is that weight which contains one gram atom of ionisable hydrogen or its chemical equivalent. For example, the equivalent weight of sulfuric acid (H_2SO_4) is 49 (98 divided by 2 because there are two replaceable hydrogen ions). A one N solution of sulfuric acid would consist of 49 grams of H_2SO_4 dissolved in enough water to make one litre of solution.

- **normal distribution**
  particle size distribution characterised by a bell-shaped or Gaussian distribution shape when plotted on a linear scale.

- **nosocomial**
  associated with a hospital or infirmary, especially used of

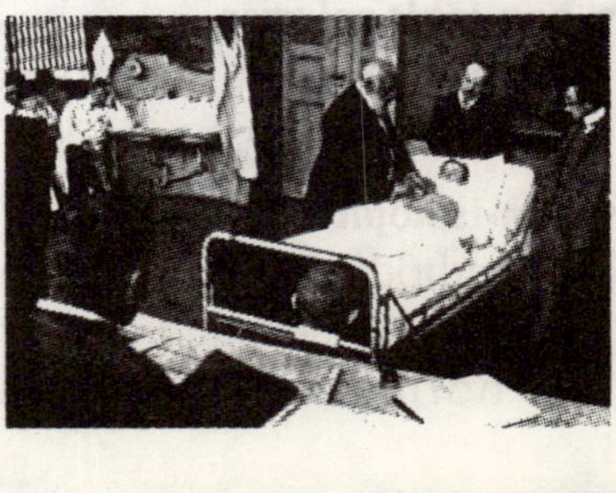

diseases that may result from treatment in such an institution. See **iatrogenic**.

- **noxious substance**
  see **harmful substance**.

- **nuclear magnetic resonance (nmr)**
  the use of magnetic fields and radio waves (instead of the X-rays employed in the CT scan) to visualise body structures and how they function. The powerful magnet, large enough to enclose the body, forces the atomic nuclei in most material such as soft tissue to align themselves with the magnetic field. Radio waves are aimed at the selected area to excite the atoms which, when the waves are stopped, emit signals that are converted to computer-generated pictures. NMR is used to detect diseases of soft tissues, heart, brain, and spinal cord. NMR is also called magnetic resonance imaging (MRI).

- **nuisance threshold**
  lowest concentration of an air pollutant that can be considered objectionable. See **odour threshold, pollutant**.

- **nutrient**
  any substance that is assimilated (taken in) by organisms and promotes growth. Nitrogen and phosphorous are nutrients which promote the growth of algae. There are other essential and trace elements which are also considered nutrients.

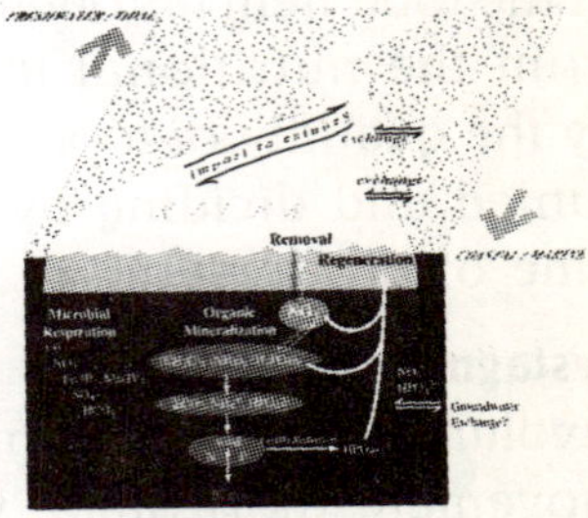

- **nutrient pollution**
  contamination of water resources by excessive inputs of nutrients. In surface waters, excess algal production is a major concern.

- **nutritional table method**
  procedure for evaluating the dietary intake of a large number of people. The accuracy of the method depends on the accuracy with which records of the food consumption can be estab-

lished and the accuracy of the nutritional tables specifying the concentration of various nutrients, vitamins, essential, and non-essential substances including pesticide residues. For each record of quantity of food consumed during a certain time period, the daily intake of the substance in question is calculated by multiplying the substance concentration in the food item (as obtained from the nutritional table) by the quantity of food consumed and dividing by the time of observation.

- **nystagmus**
  involuntary, rapid, rhythmic movement (horizontal, vertical, rotary, mixed) of the eyeball, usually caused by a disorder of the labyrynth of the inner ear or a malfunction of the central nervous system.

- **objective environment**
  actual physical, chemical, and social environment as described by objective measurements, such as noise levels in decibels and concentrations of air pollutants.

- **occupational environment**
  surrounding conditions at a workplace.

- **occupational exposure**
  experience of substances, intensities of radiation etc. or other conditions while at work.

- **occupational exposure limit (OEL)**
  regulatory level of exposure to substances, intensities of radiation etc. or other conditions , specified appropriately in relevant government legislation or related codes of practice.

- **occupational exposure standard (OES)**
  level of exposure to substances, intensities of radiation etc. or other conditions considered to represent specified good practice and a realistic criterion for the control of exposure by appropriate plant design, engineering controls, and, if necessary, the addition and use of personal protective clothing.

- **occupational hygiene**
  identification, assessment and control of physico-

chemical and biological factors in the workplace that may affect the health or well-being of those at work and in the surrounding community.

- **odour threshold**
  the minimum odour of a water sample that can just be detected after successive dilutions with odourless water. Also called threshold odour.

- **octanol-water partition coefficient (pow, kow)**
  measure of lipophilicity by determination of the equilibrium distribution between octan-1-ol and water, as used in pharmacological studies and in the assessment of environmental fate and transport of organic chemicals.

- **ocular**
  pertaining to the eye.

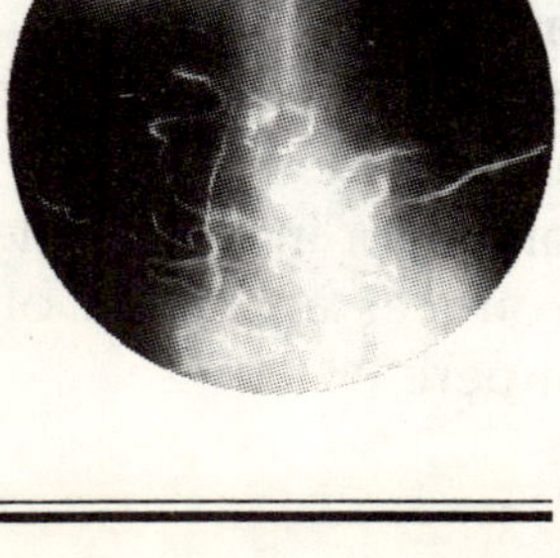

- **odds**
  ratio of the probability of occurrence of an event to that of non-occurrence, or the ratio of the probability that something is so, to the probability that it is not so.

- **odds ratio**
  quotient obtained by dividing one set of odds by another. The term 'odds' or 'odds ratio' is defined differently according to the situation under discussion. Consider the following notation for the distribution of a binary exposure and a disease in a population or a sample.

	Exposed	Nonexposed
Disease	*a*	*b*
No disease	*c*	*d*

The odds ratio (cross-product ratio) is *ad/(bc)*. The exposure-odds ratio for a set of case control data is the ratio of the odds in favour of exposure among the cases (*a/b*) to the odds in favour of exposure among non-cases (*c/d*). This reduces to *ad/(bc)*. With incident cases, unbiased subject selection, and a 'rare' disease (say, under 2 % cumulative incidence rate over the study period), *ad/(bc)* is an approximate estimate of the risk ratio. With incident cases, unbiased subject selection, and density sampling of controls, *ad/(bc)* is an estimate of the ratio of the person-time incidence rates (forces of morbidity) in the exposed and unexposed. No rarity assumption is required.

The disease-odds (rate-odds) ratio for a cohort or cross section is the ratio of the odds in favour of disease among the exposed population *(a/c)* to the odds in favour of disease among the unexposed *(b/d)*. This reduces to *ad/(bc)* and hence is equal to the exposure-odds ratio for the cohort or cross section.

The prevalence-odds refers to an odds ratio derived cross sectionally, as, for example, an odds ratio derived from studies of prevalent (rather than incident) cases. The risk-odds ratio is the ratio of the odds in favour of getting disease, if exposed, to the odds in favour of getting disease if not exposed. The odds ratio derived from a cohort study is an estimate. See **cross-product ratio, relative odds**.

- **oedema**

presence of abnormally large amounts of fluid in intercellular spaces of body tissues.

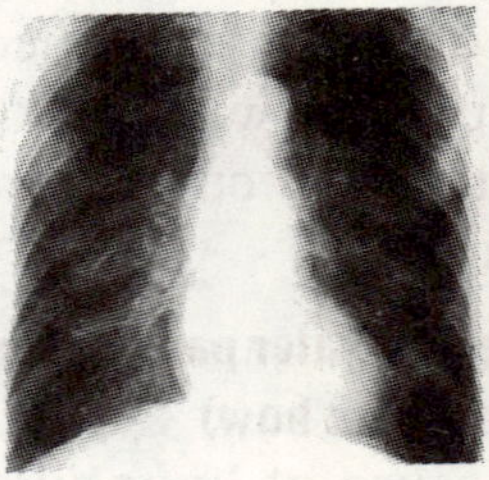

- **offset (or droop)**

the difference between the actual value and the desired value (or set point). Characteristic of proportional controllers that do not incorporate reset action.

- **off-stream uses**

water withdrawn from surface or ground water sources for use at another place.

- **ohm**

the unit of electrical resistance. The resistance of a conductor in which one volt produces a current of one ampere.

- **olfactometer**
  apparatus for testing the power of the sense of smell.

- **olfactory fatigue**
  a condition in which a person's nose, after exposure to certain odours, is no longer able to detect the odour.
- **oligotrophic**
  reservoirs and lakes which are nutrient poor and contain little aquatic plant or animal life.
- **oliguria**
  excretion of a diminished amount of urine in relation to fluid intake.
- **on site**
  on the same, or adjacent, property.
- **oncogene**
  gene that can cause neoplastic transformation of a cell. Oncogenes are slightly changed equivalents of normal genes known as proto-oncogenes. See **transformation**.

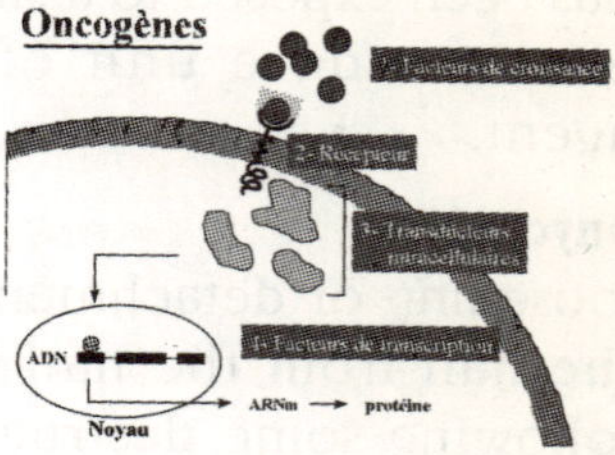

- **oncogenesis**
  production or causation of tumours.

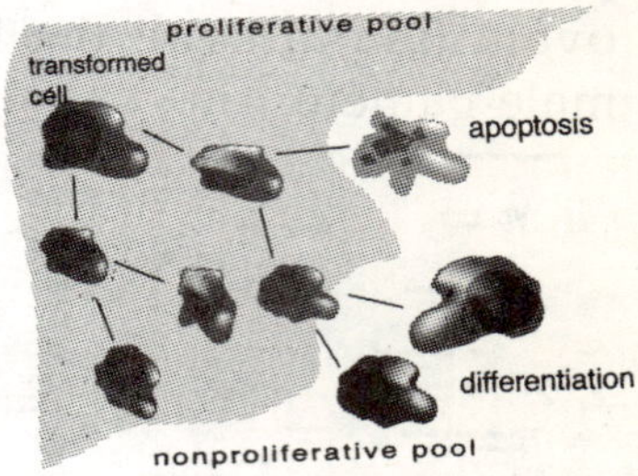

- **oncologist**
  a specialist in the treatment of cancer.
- **oncology**
  study of cancer.
- **one-hit model**
  dose-response model of the form
  $P(d) = 1 - \exp(-bd)$
  where $P(d)$ is the probability of cancer death from a continuous dose rate (d) and b is a constant. The one-hit model is based on the con-

cept that a tumour can be induced after a single susceptible target or receptor has been exposed to a single effective dose unit of an agent.

- **onycholysis**
loosening or detachment of the nail from the nail bed following some destructive process.

- **oogenesis**
process of formation of the ovum (plural ova), the female gamete.

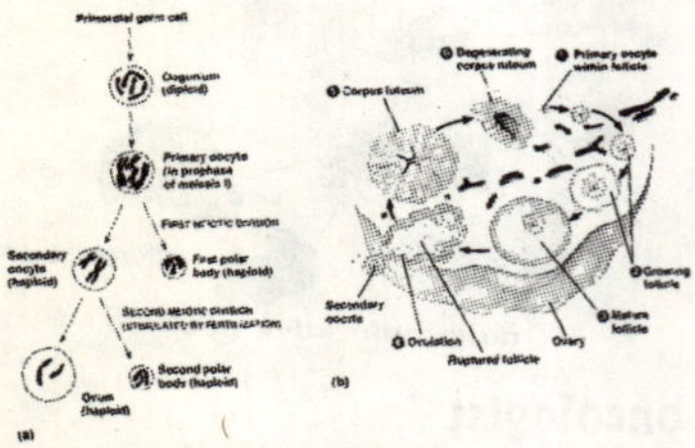

- **open-face sampler**
a filter cassette sampler with the inlet approximately the same size as the filter.

- **operation and maintenance costs**
the ongoing, repetitive costs of operating a water system, for example, employee wages and costs for treatment chemicals and periodic equipment repairs.

- **operon**
complete unit of gene ex pression and regulation, in cluding structural genes regulator gene(s) and con trol elements in DN recognised by regulato gene product(s).

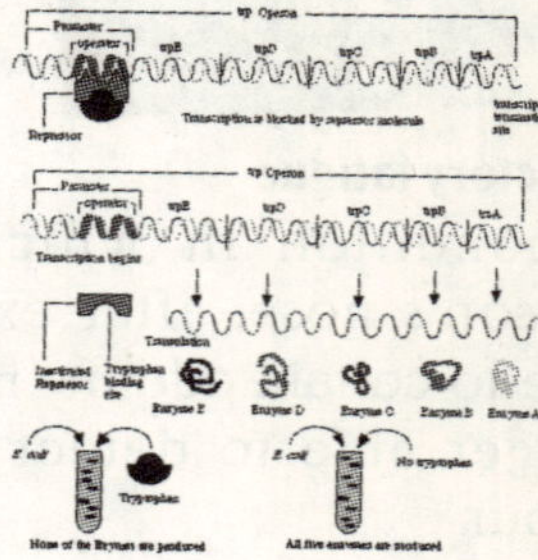

- **ophthalmic**
pertaining to the eye.

- **opiates**
drugs containing or derived from opium (a poppy juice) used to relieve pain or induce sleep. They are addictive narcotics.

- **optimal corrosion control treatment**
the corrosion control treatment that minimises the lead and copper concentrations at users' taps while insuring that the treatment does not cause the water

system to violate any national primary drinking water regulations.

- **oral glucose tolerance test**
a series of blood tests measuring plasma glucose concentration, the first after an 8 to 12 hour fast, then a liquid that contains 75 grams of glucose is ingested, after which the second blood glucose is measured at 30 minutes, the third at one hour, the fourth at two hours and the fifth at three hours. A diagnosis of diabetes is made if the venous plasma glucose level is > 200 mg/dl at 120 min. and one other blood glucose level is also > 200 mg/dl.

- **oral.**
of the mouth, through or by the mouth.

- **orchiectomy**
surgical removal of the testicles.

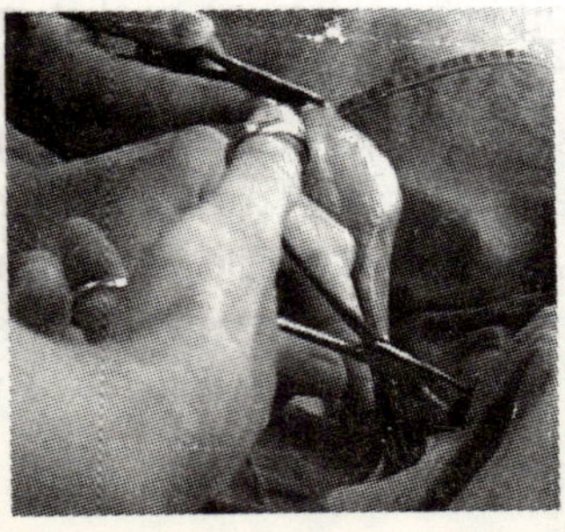

- **organ dose**
amount of a substance or physical agent (radiation) absorbed by an organ.

- **organelle**
microstructure or separated compartment within a cell that has a specialised function, for example ribosome, peroxisome, lysosome, Golgi apparatus, mitochondrion, nucleolus, nucleus.

- **organic**
generally considered as originating from plants or animals, and made primarily of carbon and hydrogen. Scientists use the term organic to mean those chemical compounds which are based on carbon.

- **organic carbon partition coefficient (koc)**
measure of the tendency for organic substances to be adsorbed by soil and sediment, expressed as:

$$K_{OC} = \frac{(\text{mg substance adsorbed})/(\text{kg organic carbon})}{(\text{mg substance dissolved})/(\text{litre of solution})}$$

The K_{OC} is substance-specific and is largely independent of soil properties.

- **organic toxic agent**
  toxic agents that are organic compounds

- **organically grown**
  food, feed crops, and livestock grown within an intentionally-diversified, self-sustaining agro-ecosystem. In practice, farmers build up nutrients in the soil using compost, agricultural wastes, and cover crops instead of synthetically derived fertilisers to increase productivity, rotate crops, weed mechanically, and reduce dramatically their dependence on the entire family of pesticides. Farmers must' be certified to characterise crops as organically grown and can only use approved natural and synthetic biochemicals, agents, and materials for three consecutive years prior to harvest. Livestock must be fed a diet that includes grains and forages that have been organically grown and cannot receive hormones, sub-therapeutic antibiotics, or other growth promoters.

- **organism**
  any living being, whether plant, mammal, bird, insect, reptile, fish, crustacean, aquatic or estuarine animal or bacterium.

Number of Living Species of All Kinds of Organisms Currently Known (According to Major Group)

ALL ORGANISMS: TOTAL SPECIES, 1,413,000

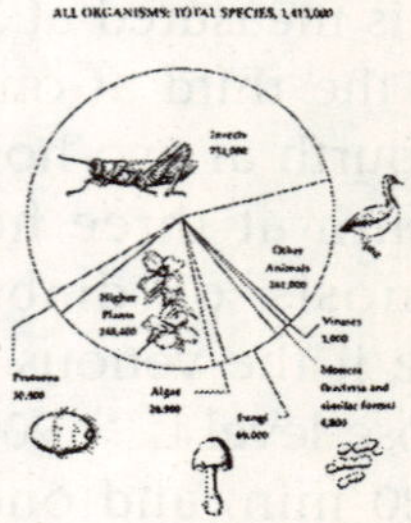

- **organoleptic**
  involving an organ, especially a sense organ as of taste, smell or sight.

- **orifice**
  an opening (hole) in a plate, wall or partition An orifice flange or plate placed in a pipe consists of a slot or a calibrated circular hole smaller than the pipe diameter. The difference in pressure in the pipe above and at the orifice may be used to determine the flow in the pipe.

- **orifice meter**
  a device used to measure flow rate in a duct by measuring the pressure drop across a calibrated constriction. Also known as a variable pressure meter, the flow rate is a function of the pressure drop across a constriction (orifice) in the flow stream and the size of the orifice.

- **orp. oxidation-reduction potential**
  the electrical potential required to transfer electrons from one compound or element (the oxidant) to another compound or element (the reductant). Used as a qualitative measure of the state of oxidation in water treatment systems.

- **orthotolidine**
  orthotolidine is a calorimetric indicator of chlorine residual. If chlorine is present, a yellow-colored compound is produced. This reagent is no longer approved for chemical analysis.

- **osmosis**
  the tendency of a fluid to pass through a permeable membrane, like the wall of a living cell, into a less concentrated solution, so as to equalise concentrations on both sides of the membrane.

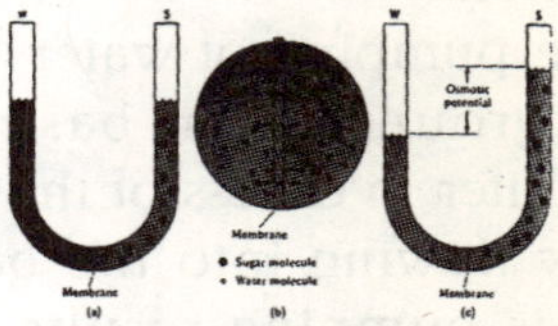

El potencial osmótico se puede poner de manifiesto si colocamos en un tubo cerrado en forma de U, en una parte agua y en la otra una disolución, por ejemplo de agua con azúcar. Ambas se encuentran separadas por una membrana semipermeable (parte a de la figura).
Esta membrana es permeable para el agua pero impermeable para las moléculas de azúcar (parte b).
La disolución atrae agua para diluirse y tratar de equiparar las concentraciones a cada lado de la membrana, creándose un potencial osmótico (parte c).

- **osteo-**
  prefix meaning pertaining to bone.

- **osteodystrophy**
  abnormal development of bone.

- **osteogenesis**
  formation or development of bone.

- **osteoporosis**
  significant decrease in bone mass with increased porosity and increased tendency to fracture.

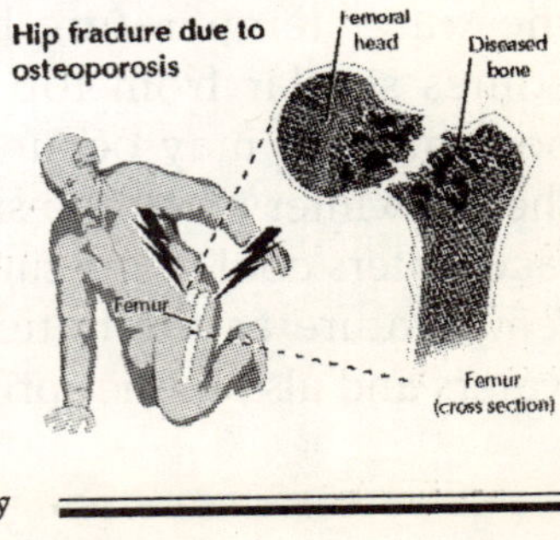

- **overall efficiency pump**
  the combined efficiency of a pump and motor together. Also called the wire-to-water efficiency.

- **overdraft**
  the pumping of water from a groundwater basin or aquifer in excess of the supply flowing into the basin. This pumping results in a depletion or 'mining' of the groundwater in the basin.

- **overflow rate**
  one of the guidelines for the design of settling tanks and clarifiers in treatment plants. Used by operators to determine if tanks and clarifiers are hydraulically (flow) over- or under-loaded. Also called surface loading. Overflow Rate (GDP/sq ft) = Flow (GPD)/Surface Area (sq ft)

- **overturn**
  the almost spontaneous mixing of all layers of water in a reservoir or lake when the water temperature becomes similar from top to bottom. This may occur in the fall/winter when the surface waters cool to the same temperature as the bottom waters and also in the spring when the surface waters warms after the ice melts.

- **ovicide**
  substance intended to kill eggs.

- **oxidant**
  a substance containing oxygen that reacts chemically with other materials to produce new substances. Oxidants are the primary ingredients in smog.

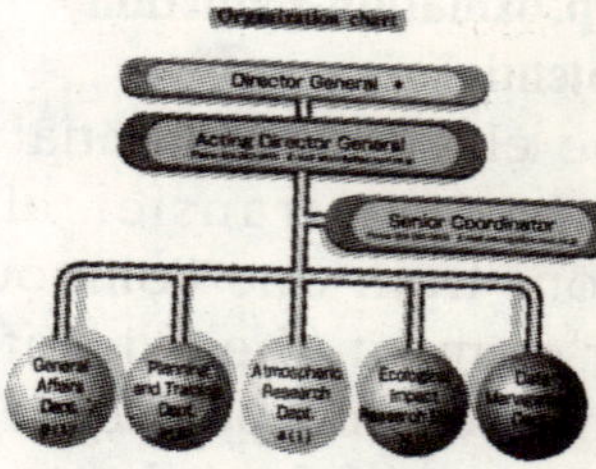

- **oxidation**
  1. the addition of oxygen which breaks down organic waste or chemicals such as cyanides, phenols, and organic sulphur compounds in sewage by bacterial and chemical means.
  2. oxygen combining with other elements.
  3. the process in chemistry whereby electrons are removed from a molecule.

- **oxidation pond**
  a man-made lake or body of water in which waste is con-

sumed by bacteria. It is used most frequently with other waste-treatment processes. An oxidation pond is basically the same as a sewage lagoon.

- **oxidation-reduction potential.**
the electrical potential required to transfer electrons from one compound or element (the oxidant) to another compound or element (the reductant). Used as a qualitative measure of the state of oxidation in water treatment systems.

- **oxidising agent**
a chemical or substance that brings about an oxidation reaction. The agent may (1) provide the oxygen to the substance being oxidised (in which case the agent has to be oxygen or contain oxygen), or (2) receive electrons being transferred from the substance undergoing oxidation. (Chlorine is a good oxidising agent for electron-transfer purposes, even though it contains no oxygen), (see **Reducing Agent**).

- **ozonation**
the application of ozone to water for disinfection or for taste and odour control.

- **ozone (O_3)**
found in two layers of the atmosphere, the stratosphere and the troposphere. In the stratosphere (the atmospheric layer beginning 7 to 10 miles above the earth's surface) ozone is a form of oxygen found naturally which provides a protective layer shielding the earth from ultraviolet radiation's harmful health effects on humans and the environment. In the troposphere (the layer extending up 7 to 10 miles from the earth's surface), ozone is a chemical oxidant and major component of photochemical smog. Ozone can seriously affect the human respiratory system. Ozone in the troposphere is produced through complex chemical reactions of nitrogen oxides, which are among the primary pollutants emitted by combustion sources.

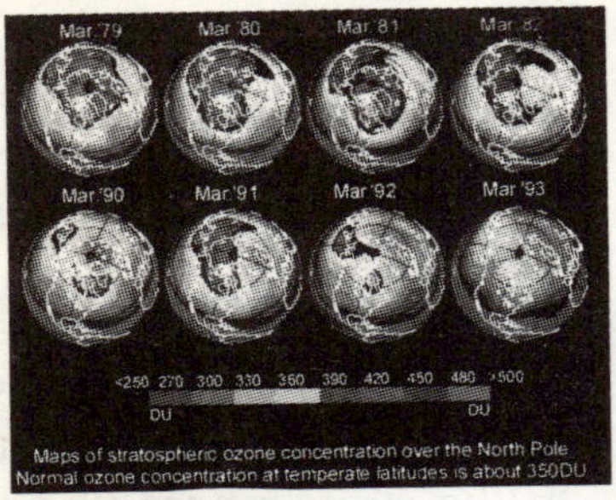

Maps of stratospheric ozone concentration over the North Pole
Normal ozone concentration at temperate latitudes is about 350DU

- **ozone depletion**
  destruction of the stratospheric ozone layer which shields the earth from ultraviolet radiation harmful to biological life. This destruction of ozone is caused by the breakdown of certain chlorine and/or bromine containing compounds (chlorofluorocarbons or halons) which break down when they reach the stratosphere and catalytically destroy ozone molecules.

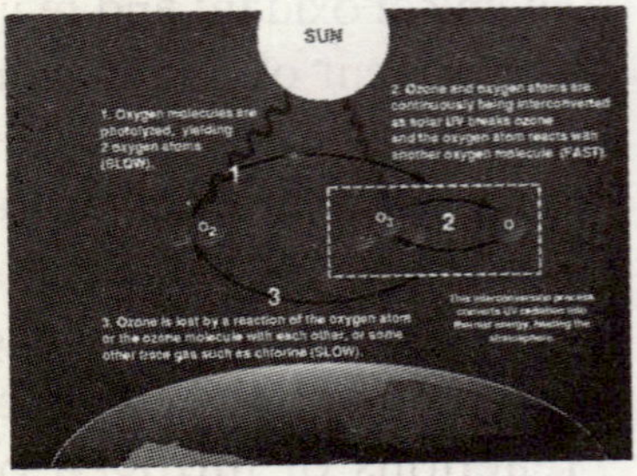

- **p.o.**
  this is the abbreviation for per os, signifying oral administration.

- **packed tower aeration**
  a method of treating water to remove volatile organic chemical (VOCs) contaminants. As water is mixed with air, VOCs move from water to air which then passes through carbon filters to trap the contaminants.

- **palatable**
  water at a desirable temperature that is free from objectionable tastes, odours, colours, and turbidity. Pleasing to the senses.

- **palpitation**
  1. unduly rapid or throbbing heartbeat that is noted by a patient; it may be regular or irregular.
  2. undue awareness by a patient of a heartbeat that is otherwise normal.

- **panel testing**
  a laboratory procedure in which a series of tests is performed on one specimen, usually related to a single condition or disease, or for differential diagnosis.

- **paraesthesia**
  abnormal sensation, a burning or prickling.

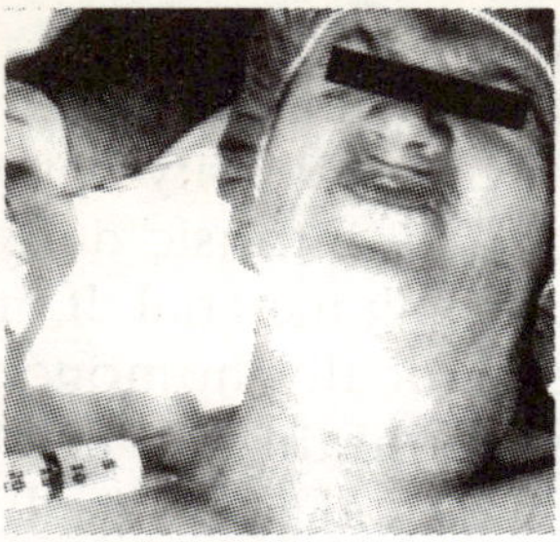

- **parakeratosis**
in medicine, this term is applied to the imperfect formation of horn cells of the epidermis.

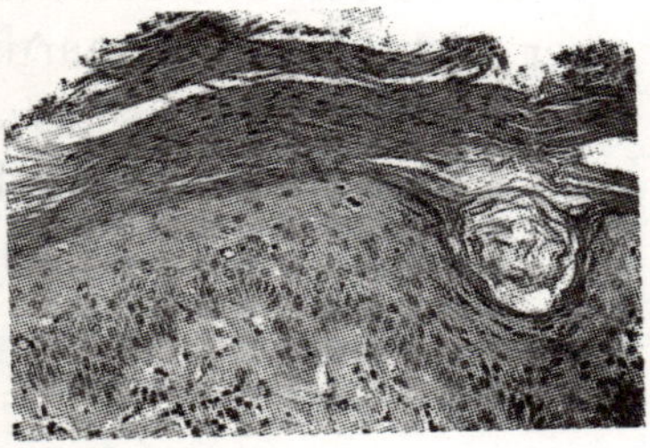

- **paralysis**
loss or impairment of motor function.

- **parametric hypothesis tests**
hypothesis test methods that assume data are normally distributed and variance is homogeneous. These tests are considered to be more powerful than nonparametric equivalents. Examples ANOVA, Dunnettis, Studenti's tests.

- **para-occupational exposure**
exposure of a worker's family to substances carried from the workplace to the home.
Exposure of visitors to substances in the workplace.

- **parasympatholytics**
producing effects resembling those caused by interruption of the parasympathetic nerve; also called anticholinergic.

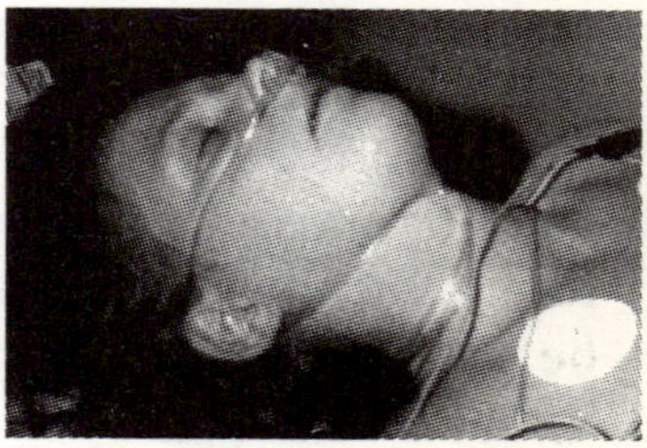

- **parenchyma (-al)**
in medicine, this term is applied to a specific or functional component of a gland or organ.

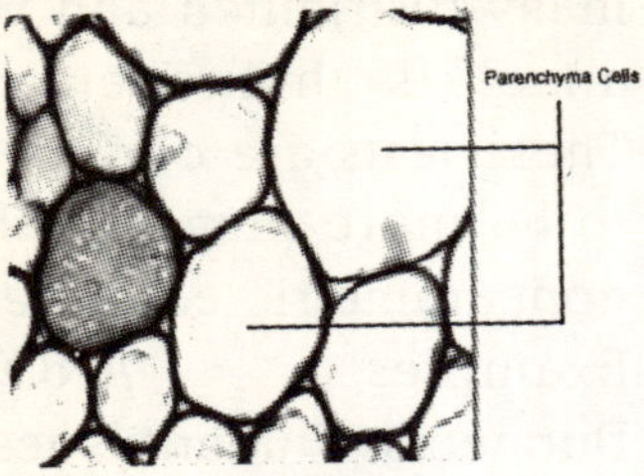

- **parenteral dosage**
  method of introducing substances into an organism avoiding the gastrointestinal tract (subcutaneously, intravenously, intramuscularly etc.).

- **paresis**
  slight or incomplete paralysis.

- **parshall flume**
  a device used to measure the flow in an open channel. The flume narrows to a throat of fixed dimensions and then expands again. The rate of flow can be calculated by measuring the difference in head (pressure) before and after at the throat of the flume.

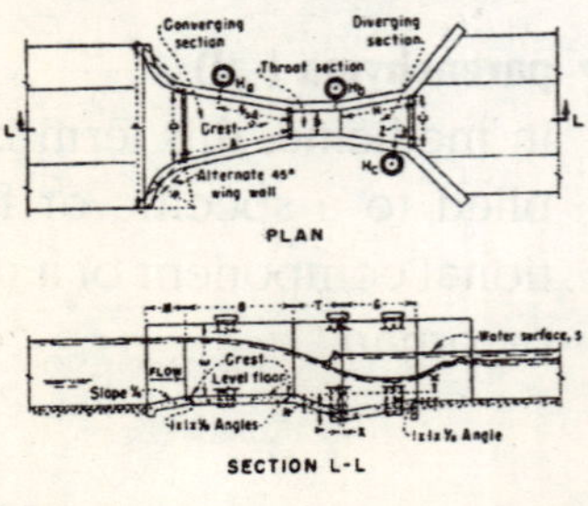

- **particle**
  a small discrete object, ofte having a density approaching the intrinsic density o the bulk material. It may b chemically homogeneou or contain a variety o chemical species. It ma consist of solid or liqui materials or both.

- **particle count**
  the results of a microscopi examination of treated water with a special 'particl counter' which classifies suspended particles by numbe and size.

- **particulate**
  1. small, separate pieces o an airborne material. Dusts fumes, smokes, mists, an fogs are examples. Generally, anything that is not a fibre and has an aspect ratio of 3 to 1.

  2. a very small solid suspended in water which ca vary widely in size, shape density, and electrica charge. Colloidal and dispersed particulates are artificially gathered together b the processes of coagulatio and flocculation.

- **partition coefficient**
  1. ratio of the distribution of a substance between two phases when the heterogeneous system (of two phases) is in equilibrium. The ratio of concentrations (or, strictly speaking, activities) of the same molecular species in the two phases is constant at constant temperature. The partition coefficients most frequently used in acute toxicology are lipid/water and octan-1-ol/water distributions.

- **RT lgPOW.**
  2. a measure of the extent to which a pesticide is divided between the soil and water phases.

- **parts per billion (ppb)**
  one ppb is comparable to one kernel of corn in a filled, 45-foot silo, 16 feet in diameter.

- **parts per million (ppm)**
  one ppm is comparable to one drop of gasoline in a thank full of gas (full-size car).

- **parts per trillion (ppt)**
  one ppt is comparable to one drop in a swimming pool covering the area of a football field 43 ft. deep.

- **pascal.**
  the pressure or stress of one newton per square meter. (Abbreviated Pa) Pa = 6.895 kN/sq m = 0.0703 kg/sq cm

- **passive smoking**
  inhalation of sidestream smoke by people who do not smoke themselves. See **sidestream smoke**.

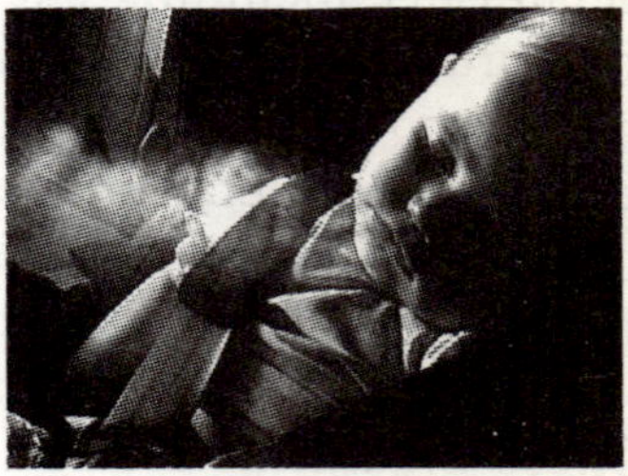

- **pathogen**
  micro-organisms typically found in the intestinal tracts of mammals that can cause disease in other organisms or in humans, animals and plants. They may be bacteria, viruses, or parasites and are found in sewage, in runoff from animal farms or rural areas populated with domestic and/or wild animals, and in water used for swimming. Fish and shellfish contaminated by pathogens, or the contaminated

water itself, can cause serious illnesses.

- **pathogen interference**
  an increase in the incidence of false positive or false negative results caused by organisms pathogenic to the test species and which are present in the test sample(s).
- **pathogenic organisms**
  organisms, including bacteria, viruses or cysts, capable of causing diseases (typhoid, cholera, dysentery) in a host (such as a person). There are many types of organisms which do NOT cause disease. These organisms are called non-pathogenic.

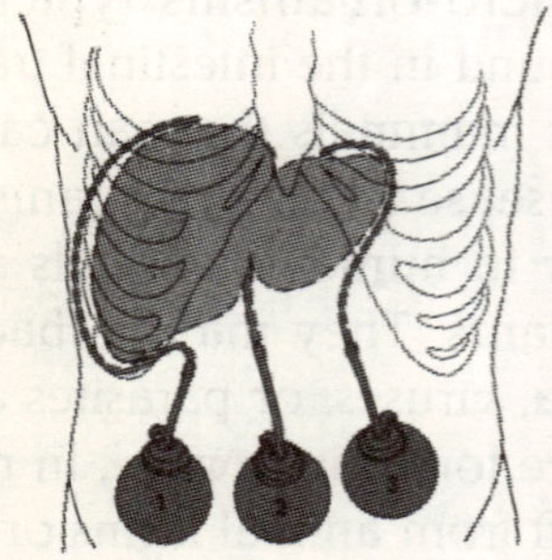

- **pathology**
  the study of disease.
- **PCP**
  see **phencyclidine hydrochloride**.
- **peak daily average concentration of an air pollutant**
  see **maximum average daily concentration of an atmospheric pollutant**.
- **pelvic inflammatory disease (PID)**
  ascending infection from the vagina or cervix to the uterus, fallopian tubes and broad ligaments: most often caused by bacterial infections of chlamydia and gonorrhoea.
- **pelvic node dissection**
  removal of possible cancer carrying lymph nodes near the prostate or their evaluation.
- **perceived environment or risk**
  see **subjective environment, risk perception**
- **percent saturation**
  the amount of a substance that is dissolved in a solution compared with the amount that could be dissolved in the solution, expressed as a percent. Amount of substance that is dissolved x 100% Amount that could be dissolved in solution

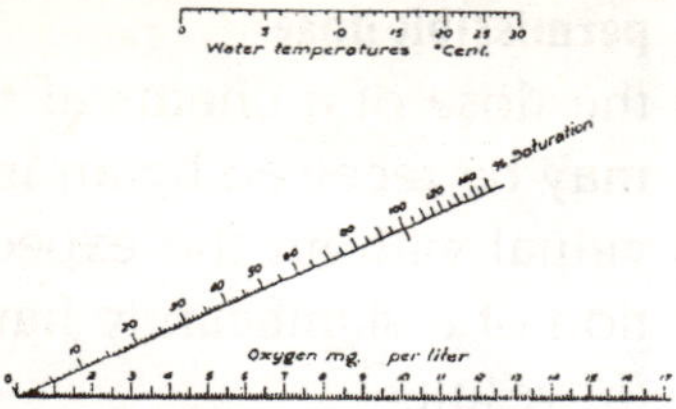

Level of oxygen saturation chart

- **percolating water**
  water that passes through soil or rocks under the force of gravity.

- **percolation**
  downward flow or filtering of water through pores or spaces in rock or soil.

- **percutaneous**
  through the skin following application on the skin.

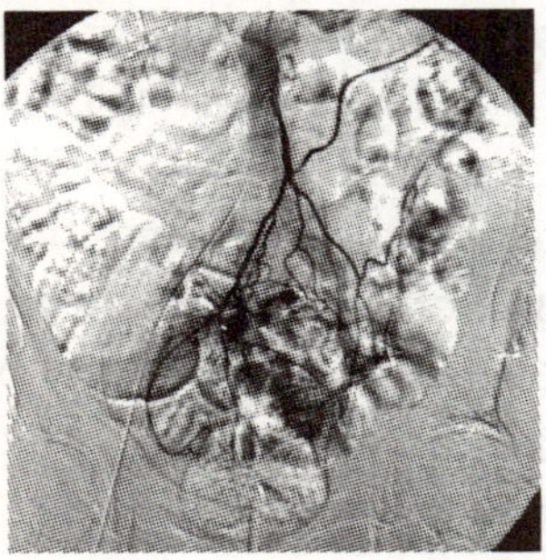

- **performance evaluation sample**
  a reference sample provided to a laboratory for the purpose of demonstrating that the laboratory can successfully analyse the sample within limits of performance specified by the Agency. The true value of the concentration of the reference material is unknown to the laboratory at the time of the analysis.

- **perinatal**
  relating to the period shortly before and after birth- from the twentieth to the twenty-ninth week of gestation to one to four weeks after birth.

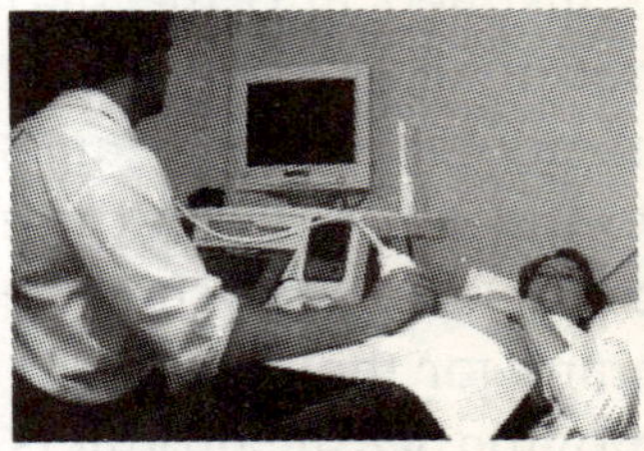

- **periodic chart (periodic table)**
  an arrangement of the elements in order of increasing atomic number that illustrates the repetition (or periodicity) of key characteristics.

- **periphyton**
  microscopic plants and animals that are firmly attached to solid surfaces under water such as rocks, logs, pilings and other structures.

- **peritoneal dialysis**
  method of artificial detoxification in which a toxic sub-

stance from the body is transferred into liquid that

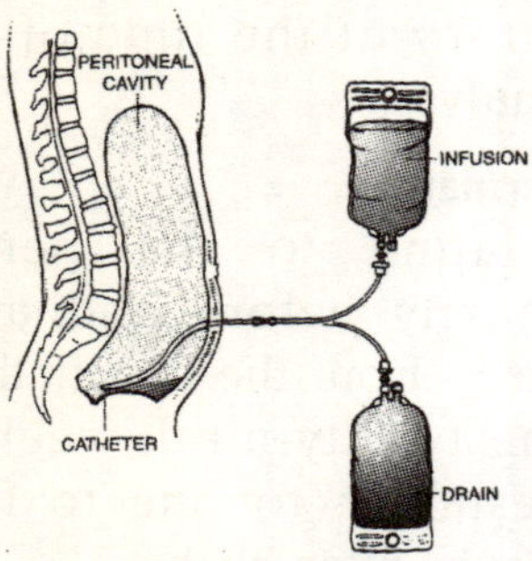

is instilled into the peritoneum. Thus, the employment of the peritoneum surrounding the abdominal cavity as a dialysing membrane for the purpose of removing waste products or toxins accumulated as a result of renal failure.

- **permeability**
  the property of permitting liquids or gases to pass through. A highly permeable soil, such as sand, allows a liquid to pass through quickly. Clay has a low permeability

- **permeate**
  to penetrate and pass through, as water penetrates and passes through soil and other porous materials.

- **permissible dose**
  the dose of a chemical that may be received by an individual without the expectation of a significantly harmful result.

- **permissible exposure limit (PEL)**
  concentration that must not be exceeded during any 8-hour work shift of a 40h working week. See **maximum allowable concentration, threshold limit value, time weighted average concentration (TWAC), exposure limit**

- **peroxisome**
  in cell biology, this is the organelle present in the cytoplasm of eukaryotic cells and characterised by its content of catalase and other oxidative enzymes such as peroxidase.

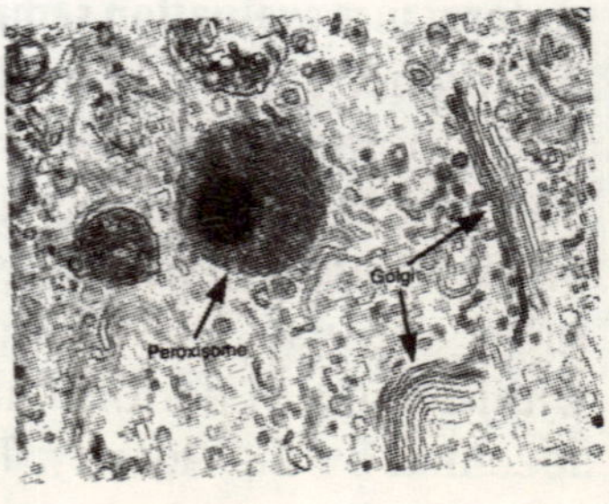

- **persistence**
the quality of remaining for a long period of time (such as in the environment or the body). Persistent chemicals (such as DDT and PCBs) are not easily broken down.

- **person**
an individual, corporation, company, association, partnership, municipality, or State or tribal agency.

- **personal monitoring**
type of environmental monitoring in which an individual's exposure to a substance is measured and evaluated: this is normally carried out using a personal sampler. See **monitoring**.

- **personal protective device (PPD)**
see personal protective equipment (PPE), individual protective device (IPD)

- **personal protective equipment (PPE)**
equipment (clothing, gloves, hard hat, respirator and so on) worn by an individual to prevent exposure to a potentially toxic substance. See **individual protective device (IPD), personal protective device (PPD)**.

- **personal sampler**
compact, portable instrument for individual air sampling, measuring, or both, the content of a harmful substance in the respiration zone of a working person. See **individual monitor**.

- **pest**
organism that may harm public health, that attacks food and other materials essential to mankind, or otherwise affects human beings adversely.

- **pesticide**
  strictly a substance intended to kill pests: in common usage, any substance used for controlling, preventing, or destroying animal, microbiological or plant pests. See **fungicide, herbicide, insecticide**

- **pesticide residue**
  pesticide residue is any substance or mixture of substances in food for man or animals resulting from the use of a pesticide and includes any specified derivatives, such as degradation and conversion products, metabolites, reaction products and impurities considered to be of toxicological significance.

- **PET scan**
  see **positron emission tomography.**

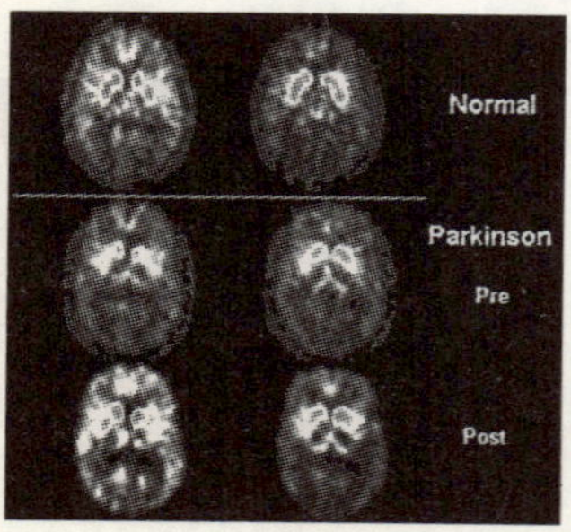

- **petroleum derivatives**
  chemicals formed when gasoline breaks down in contact with ground water.

- **pH (pronounce as separate letters)**
  1. the measure of acidity or alkalinity of a chemical solution, from 0 to 14. Anything neutral, for example has a pH of 7. Acids have a pH less than 7, bases (alkaline) greater than 7.
  2. pH is an expression of the intensity of the basic or acid condition of a liquid. Mathematically, pH is the logarithm (base 10) of the reciprocal of the hydrogen ion concentration, [H+]. pH= Log (I/[H+]) The pH may range from 0 to 14, where 0 is most acid, 14 most basic, and 7 neutral. Natural waters usually have a pH between 6.5 and 8.5.

- **phagocytosis**
  engulfing and digestion of micro-organisms, other cells, and foreign particles by cells such as phagocytes. See **macrophage**

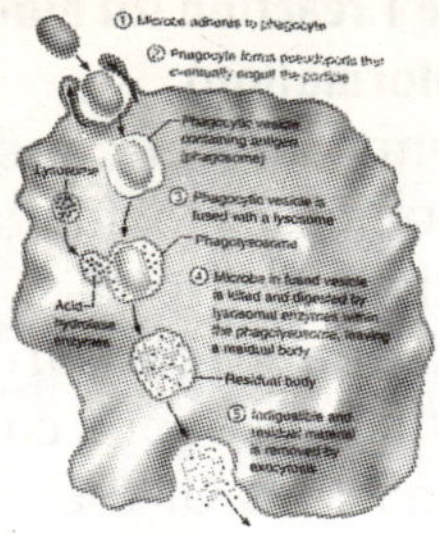

- **pharmaceuticals**
drugs, medical products, medicines, or medicaments.

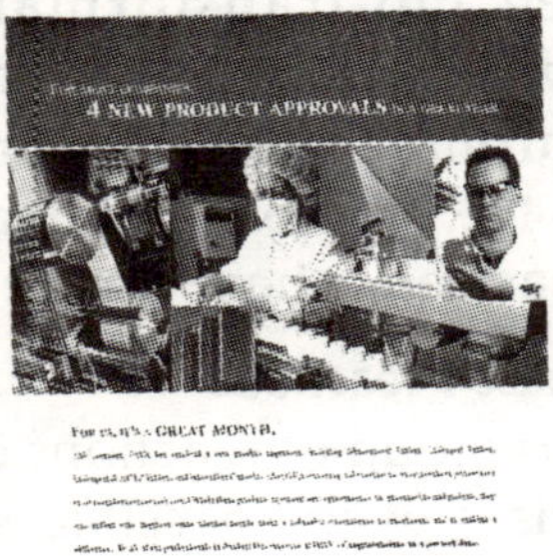

- **pharmacodynamics**
1. process of interaction of pharmacologically active substances with target sites,

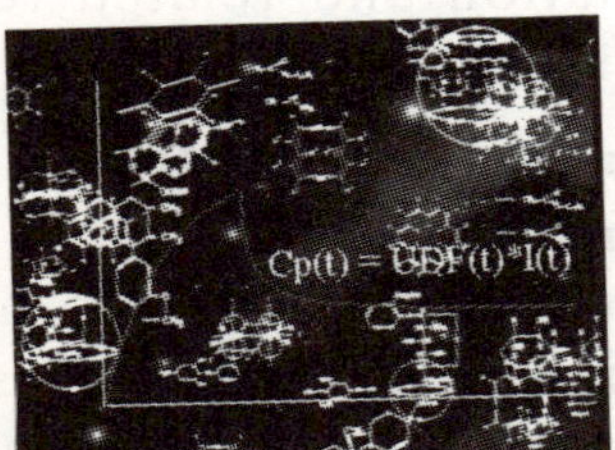

and the biochemical and physiological consequences leading to therapeutic or adverse effects. See **adverse effect, target, toxicodynamics**

2. broadly, this is the science concerned with the study of the way in which xenobiotics exert their effects on living organisms. Such a study aims to define the fundamental physicochemical processes which lead to the biological effect observed. Synonym: toxicodynamics.

- **pharmacogenetics**
study of the influence of hereditary factors on the effects of drugs on individual organisms. See **toxicogenetics, ecogenetics polymorphism**

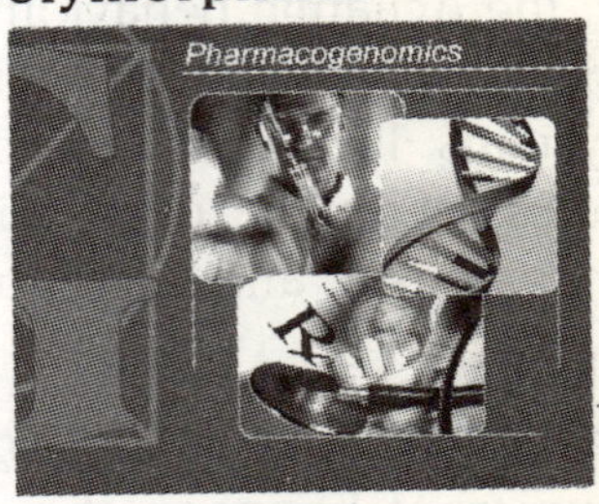

- **pharmacokinetics**
1. process of the uptake of drugs by the body, the biotransformation they undergo, the distribution of the drugs and their metabolites in the tissues, and the elimination of the drugs and their metabolites from the body. Both the amounts and the concentrations of the drugs

and their metabolites are studied. The term has essentially the same meaning as toxicokinetics, but the latter term should be restricted to the study of substances other than drugs. See **chemobiokinetics, toxicokinetics , bio-transformation, pharmacokinetics**

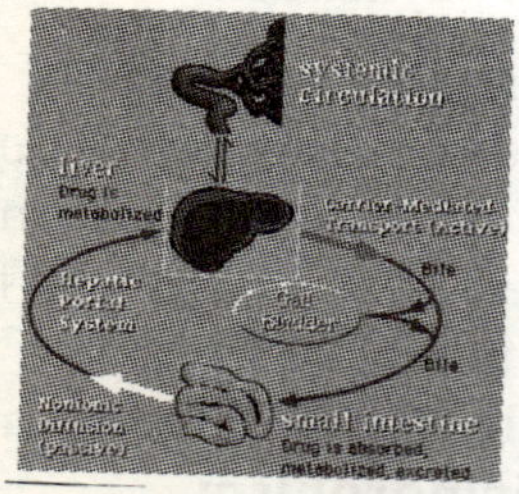

2. the dynamic behaviour of chemicals inside biological systems. It includes the processes of uptake, distribution, metabolism, and excretion.

- **pharynx**
  throat, the part of the digestive tract between the oesophagus below and the mouth and nasal cavities above and in front.

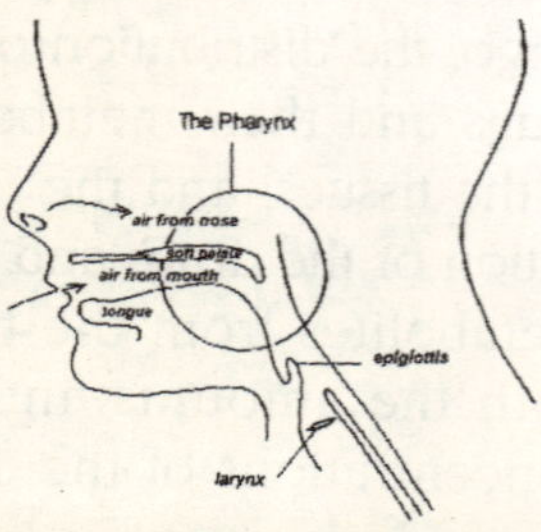

- **phase 1 reaction (of bio-transformation)**
  enzymic modification of a substance by oxidation, reduction, hydrolysis, hydration, dehydrochlorination or other reactions catalysed by enzymes of the cytosol, of the endoplasmic reticulum (microsomal enzymes) or of other cell organelles. See **bio-transformation, cytochrome P-420, cytochrome P-448, cytochrome P-450, microsome, phase 2 reaction, phase 3 reaction**

- **phase 1 reactions**
  this group of reactions comprises every possible stage in the enzymic modification of a xenobiotic by oxidation, reduction, hydrolysis, hydroxylation, dehydrochlorination and related reactions.

- **phase 2 reaction (of bio-transformation)**
  binding of a substance, or its metabolites from a phase 1 reaction, with endogenous molecules (conjugation), making more water-soluble derivatives that may be excreted in the urine or bile. See **bio-transformation,**

conjugate, phase 1 reaction, phase 3 reaction.

- **phase 3 reaction (of bio-transformation)**
  further metabolism of conjugated metabolites produced by phase 2 reactions: it may result in the production of toxic derivatives. See **bio-transformation,**

- **conjugate, phase 1 reaction, phase 2 reaction**

- **phencyclidine hydrochloride (PCP)**
  an anesthetic agent used in veterinary medicine. Also an illegal hallucinogenic street drug, 'angel dust'. Its presence may be detected by urine testing.

- **phenolic compounds**
  organic compounds that are derivatives of benzene.

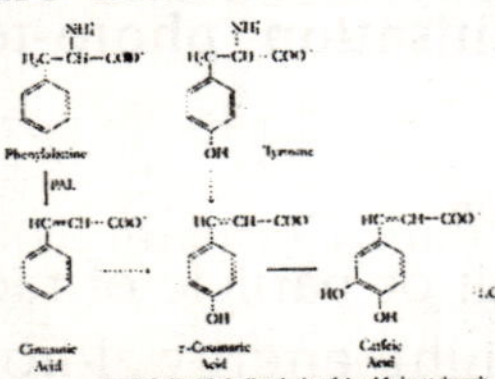

Phenolic building blocks. Deamination of phenylalanine to cinnamic acid, followed by hydroxylation to p-coumaric acid, is the first step in phenolic biosynthesis. In members of the grass family, tyrosine can be deaminated directly to p-coumaric acid. PAL = phenylalanine ammonia lyase.

- **phenolphthalein alkalinity**
  the alkalinity in a water sample measured by the amount of standard acid required to lower the pH to a level of 8.3, as indicated by the change in color of phenolphthalein from pink to clear. Phenolphthalein alkalinity is expressed as milligrams per litre equivalent calcium carbonate.

- **phenotype**
  the observable structural and functional characteristics of an organism determined by its genotype and modulated by its environment. See **genotype**

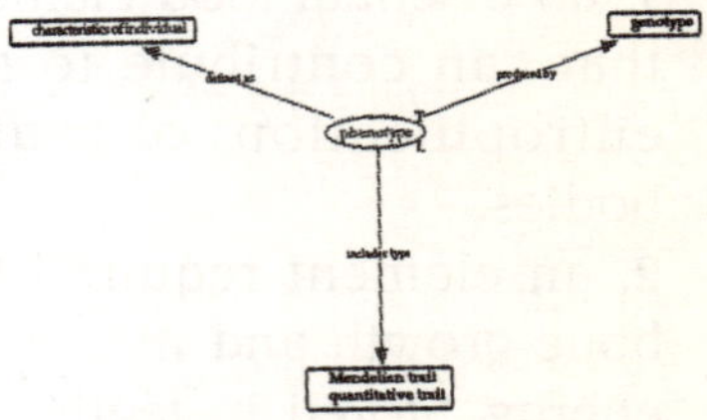

- **phenytoin**
  an anticonvulsant drug used to treat seizure disorders including epilepsy, as well as digitalis-induced heartbeat rhythm disturbances. An overdose can have toxic effects. A blood test can determine the phenytoin level to aid in correct dosing and avoid toxicity.

- **pheromone**

substance used in olfactory communication between organisms of the same species eliciting a change in sexual or social behaviour. See ectohormone, feromone

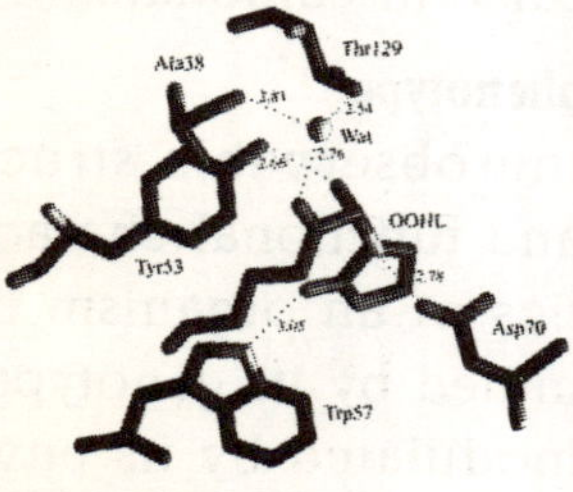

- **phosphorus (P)**

1. an essential food element that can contribute to the eutrophication of water bodies.

2. an element required for bone growth and metabolic energy. Stored in teeth and bones, mainly in combination with calcium. Phosphorus blood levels are affected by parathyroid, blood, and bone diseases.

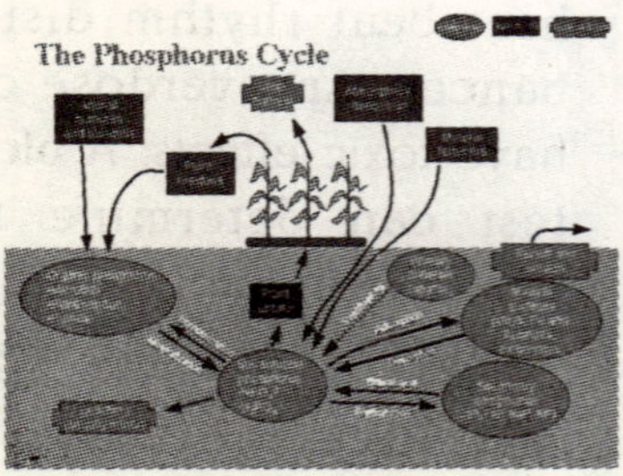

- **photochemical oxidants**

air pollutants formed by the action of sunlight on oxides of nitrogen and hydrocarbons.

- **photochemical smog**

air pollution caused by not one pollutant but by chemical reactions of various pollutants emitted from different sources.

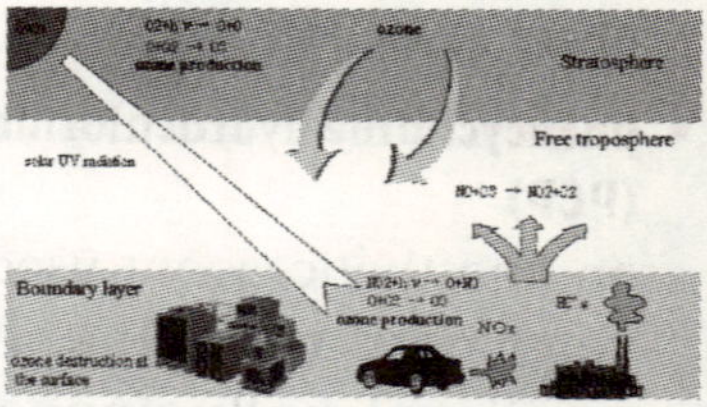

- **photo-irritation**

inflammation of the skin caused by exposure to light, especially that due to metabolites formed in the skin by photolysis. See **photosensitisation**, **photo-toxicity**.

- **photon**

a unit or particle of radiant or light energy. Photons have no electric charge. Gamma and X-rays may be referred to as photons.

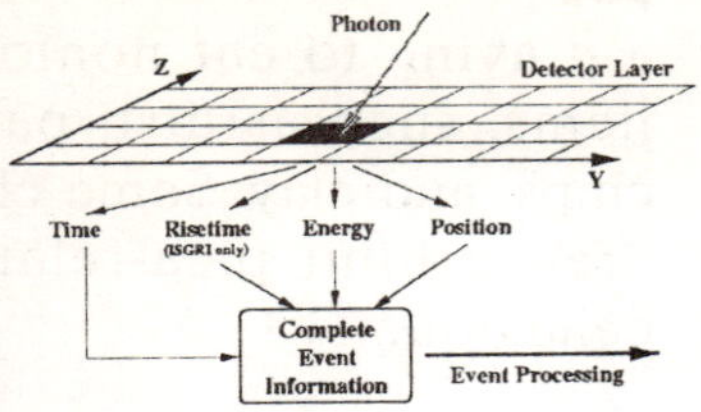

- **photo-oxidant**
  substance able to cause oxidation when exposed to light of the appropriate wavelength.

- **photophobia**
  abnormal visual intolerance of light.

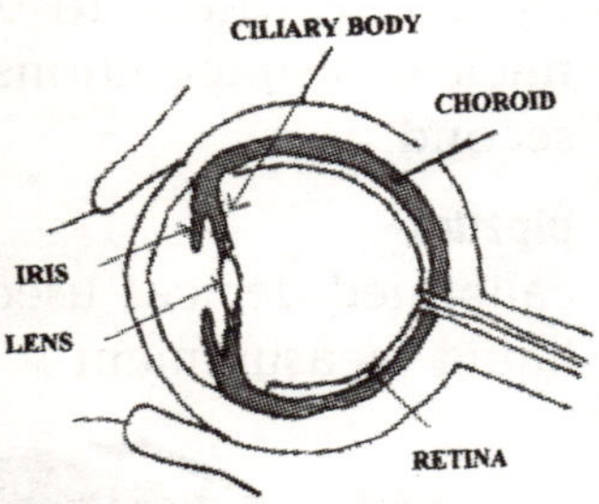

- **photo-sensitisation**
  allergic reaction due to a metabolite formed by the influence of light.

- **photosynthesis**
  the manufacture by plants of carbohydrates and oxygen from carbon dioxide and water in the presence of chlorophyll, using sunlight as an energy source.

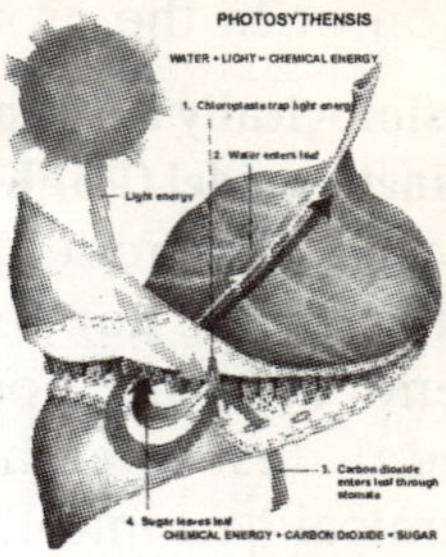

- **phototoxicity**
  adverse effects produced by exposure to light energy, especially those produced in the skin.

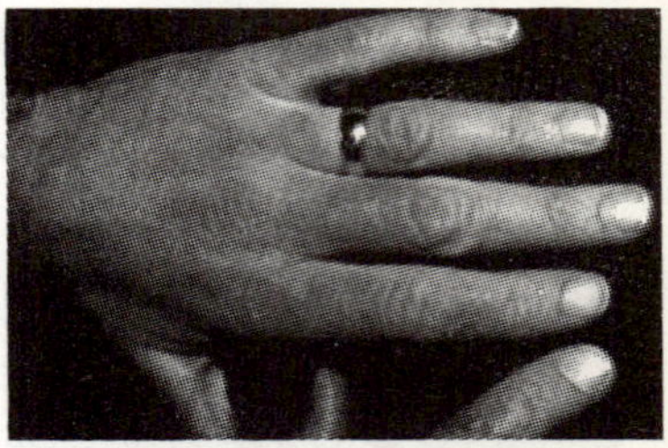

- **physical and chemical treatment**
  processes generally used in large-scale waste-water treatment facilities. Physical processes may involve air-stripping or filtration. Chemical treatment includes coagulation, chlorination, or ozone addition. The term can also refer to treatment processes, treatment of toxic materials in surface waters and ground waters, oil spills, and some methods of dealing with hazardous materi-

als on or in the ground.

- **physiologically based pharmacokinetic model (PBPK model)**
  a computer model that describes what happens to a chemical in the body. This model describes how the chemical gets into the body, where it goes in the body, how it is changed by the body, and how it leaves the body.

- **phytoplankton**
  small, usually microscopic plants (such as algae), found in lakes, reservoirs, and other bodies of water.

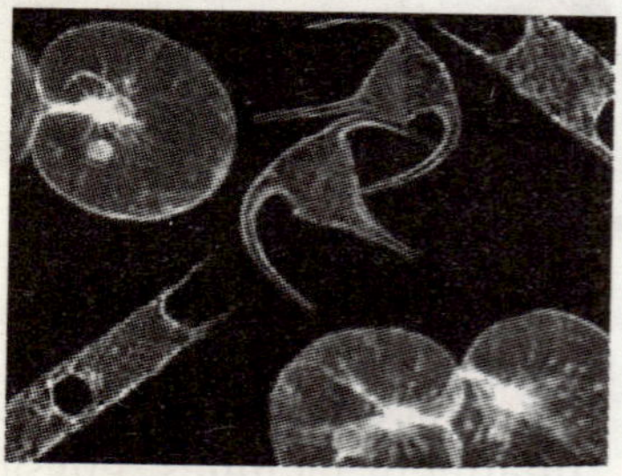

- **phytotoxic**
  poisonous to plants, inhibiting plant growth.

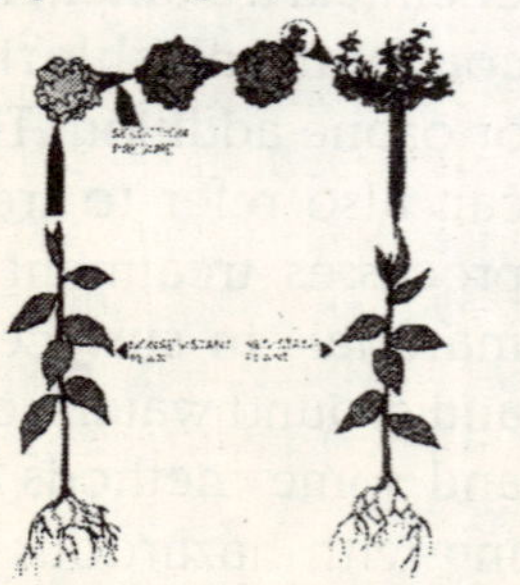

- **pica**
  a craving to eat nonfooc items, such as dirt, pain chips, and clay. Some chil dren exhibit pica-relatec behaviour.

- **pico-**
  a prefix used in the metric system and other scientific systems of measuremen which means 10^{-12} or 0.000000000001

- **picocurie (pci)**
  a measure of radioactivity. One picocurie of radioactivity is equivalent to 0.037 nuclear disintegrations per second.

- **pipettes**
  calibrated devices used for liquid measurement.

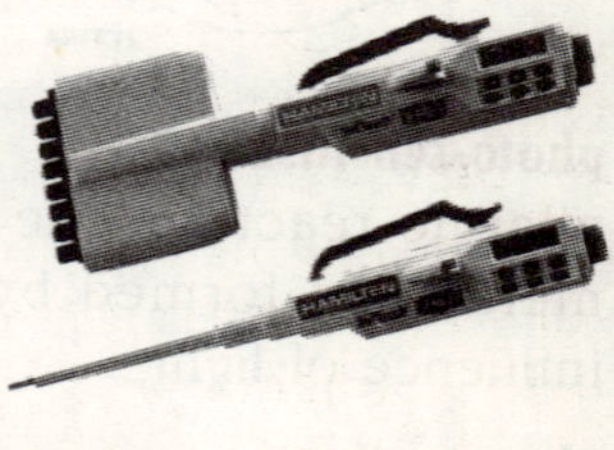

- **piscicide**
  substance intended to kill fish.

- **pituitary**
  a bi-lobed endocrine gland located at the base of the brain that secretes several

critical hormones for the regulation of other endocrine glands, including the secretion of TSH for stimulation of the thyroid. The pituitary is regulated by the neurohormones from the hypothalamus.

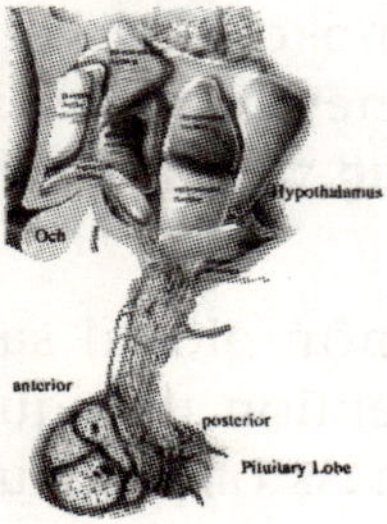

- **plan view**
  a diagram or photo showing a facility as it would appear when looking down on top of it.

- **plankton**
  small, usually microscopic, plants (phytoplankton) and animals (zooplankton) in aquatic systems.

- **plasma**
  1. fluid component of blood in which the blood cells and platelets are suspended. See **blood plasma**.
  2. fluid component of semen produced by the accessory glands, the seminal vesicles, the prostate, and the bulbourethral glands.
  3. cell substance outside the nucleus. See **cytoplasm**.

- **plasma half-life**
  see **elimination half-life**.

- **plasmapheresis**
  removal of blood from the body and centrifuging it to obtain plasma and packed red blood cells: the blood cells are resuspended in a physiologically compatible solution (usually type-specific fresh frozen plasma or albumin) and returned to the donor or injected into a patient who requires blood cells rather than whole blood.

- **plasmid**
  autonomous self-replicating extra-chromosomal circular DNA molecule.

Plasmid Insertion

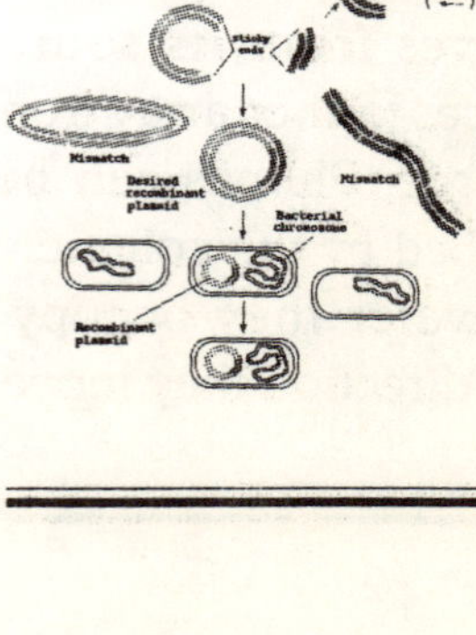

• **pleura**
lining of the lung.

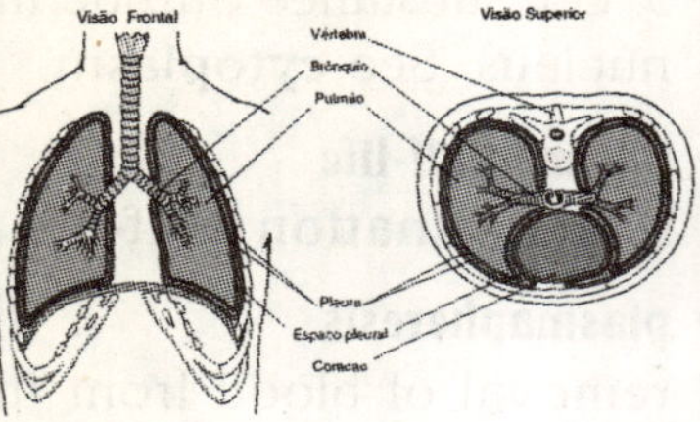

• **ploidy**
term indicating the number of sets of chromosomes present in an organism. See **diploid, haploid**.

• **plug flow**
a type of flow that occurs in tanks, basins or reactors when a slug of water moves through a tank without ever dispersing or mixing with the rest of the water flowing through the tank.

• **plumbism**
chronic poisoning caused by absorption of lead or lead salts.

• **plume**
a volume of a substance that moves from its source to places farther away from the source. Plumes can be described by the volume of air or water they occupy and the direction they move. For

example, a plume can be a column of smoke from a chimney or a substance moving with groundwater.

• **PM_{10}**
common global sampling convention that quantifies the mass of particulate air pollution in the environment. The PM_{10} convention collect particles in the ulrafine range. Most PM_{10} is derived from combustion sources.

• **pneumoconiosis**
usually fibrosis of the lungs that develops owing to (prolonged) inhalation of inorganic or organic dusts. Cause-specific types of pneumoconiosis:

· anthracosis: From coal dust.

· asbestosis: From asbestos dust.

· byssinosis: From cotton dust.

· siderosis: From iron dust.

silicosis: From silica dust.
stannosis: From tin dust.

- **point mutation**
  reaction that changes a single base pair in DNA.

- **point of disinfectant application**
  the point where disinfectant is applied and water downstream of that point is not subject to recontamination by surface water runoff.

- **point of exposure**
  the place where someone can come into contact with a substance present in the environment (see **exposure pathway**).

- **point source**
  1. single emission source in a defined location. See **area source**
  2. a stationary location or fixed facility such as an industry or municipality that discharges pollutants into air or surface water through pipes, ditches, lagoons, wells, or stacks, a single identifiable source such as a ship or a mine.

- **point-of-entry treatment device**
  a treatment device applied to the drinking water entering a house or building for the purpose of reducing contaminants in the drinking water distributed throughout the house or building

- **point-of-use treatment device**
  a treatment device applied to a single tap used for the purpose of reducing contaminants in drinking water at that one tap.

- **poison**
  substance that, taken into or formed within the organism, impairs the health of the organism and may kill it. See **toxic substance**

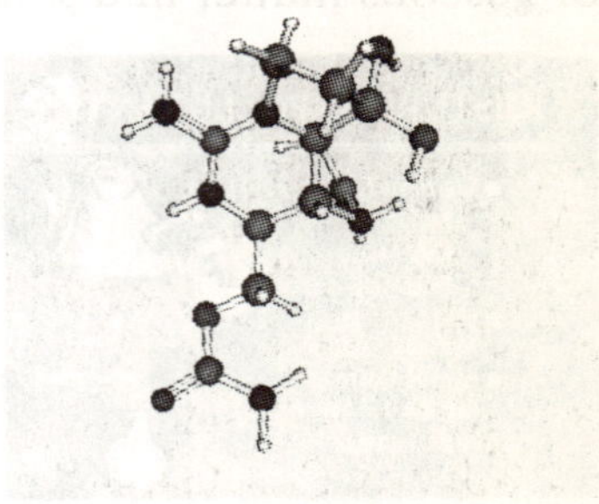

- **poison-bearing**
  containing a poison.

• **poisoning**
morbid condition produced by a poison See

• **poisson distribution**
mathematical function relating the number of particles in a given volume element to the average concentration of randomly distributed particles in the entire volume.

• **pole shader**
a copper bar circling the laminated iron core inside the coil of a magnetic starter.

• **pollutant**
any undesirable solid, liquid or gaseous matter in a solid,

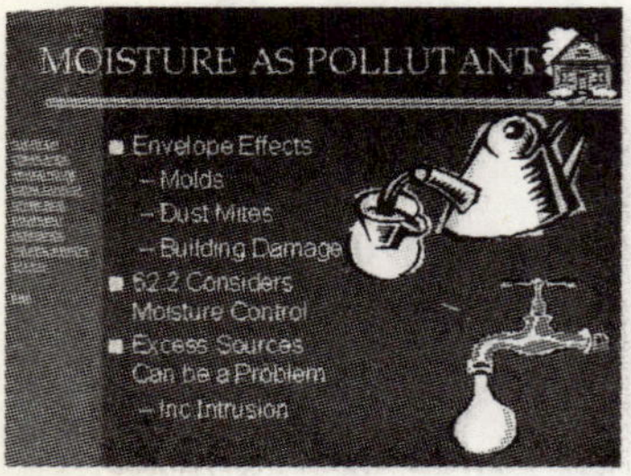

liquid or gaseous environmental medium: 'undesirability' is often concentration-dependent, low concentrations of most substances being tolerable or even essential in many cases. For the meaning of 'undesirable' in air pollution contexts, see '**pollution**'. A primary pollutant is one emitted into the atmosphere, water, sediments or soil from an identifiable source. A secondary pollutant is a pollutant formed by chemical reaction in the atmosphere, water, sediments, or soil. See **contaminant, pollution,secondary pollutant.**

• **pollution**
introduction of pollutants into a solid, liquid, or gaseous environmental medium, the presence of pollutants in a solid, liquid, or gaseous environmental medium, or any undesirable

modification of the composition of a solid, liquid or gaseous environmental medium. In the context of air pollution, an undesirable modification is one that has injurious or deleterious effects. See **contaminant, pollutant**.

• **pollution prevention**
actively identifying equipment, processes, and activities which generate excessive wastes or use toxic chemicals and then making substitutions, alterations, or product improvements. Conserving energy and minimising wastes are pollution prevention concepts used in manufacturing, sustainable agriculture, recycling, and clean air/ clean water technologies.

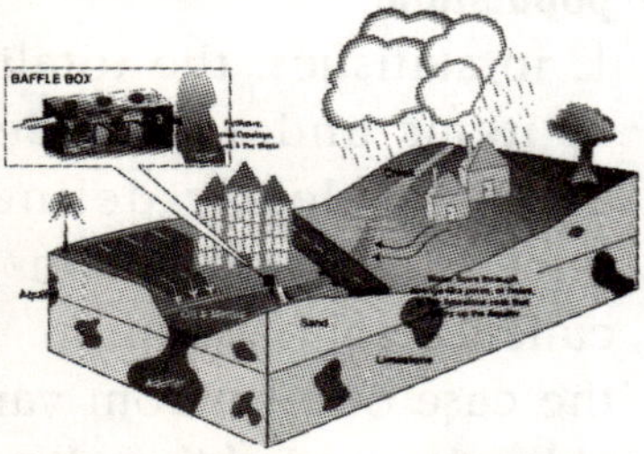

• **polychlorinated biphenyls (PCBs)**
a group of toxic, persistent chemicals used in electrical transformers and capacitors for insulating purposes, and in gas pipeline systems as a lubricant.

• **polyclonal antibody**
antibody produced by a number of different cell types. See **antibody, monoclonal antibody**.

• **polydipsia**
a chronic excessive intake of water; a characteristic of uncontrolled diabetes mellitus.

• **polyelectrolyte**
a high-molecular-weight (relatively heavy) substance having points of positive or negative electrical charges that is formed by either natural or man-made processes. Natural polyelectrolytes may be of biological origin or derived from starch products and cellulose derivatives. Man-made polyelectrolytes consist of simple substances that have been made into complex substances of high molecular weight. Used with other chemical coagulants to aid in binding small suspended particles to larger chemical flocs for their removal from water. Often called a polymer.

- **polymer**
a chemical formed by the union of many mono-mers (a molecule of low molecular weight). Polymers are used with other chemical coagulants to aid in binding small suspended particles to larger chemical flocs for their removal from water. All polyelectrolytes are polymers, but not all polymers are polyelectrolytes.

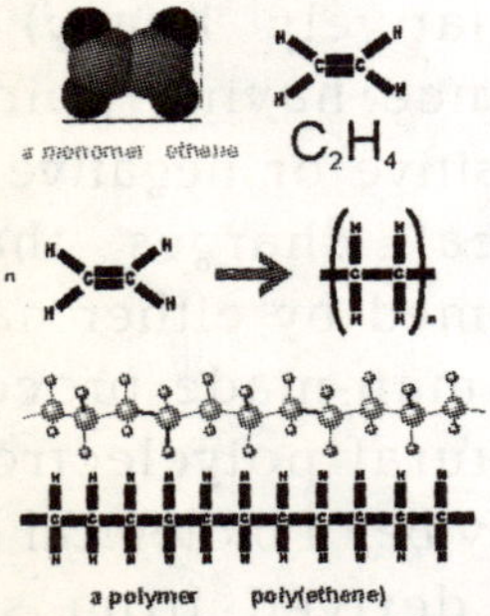

- **polymorphism (polymorphia) in metabolism**
interindividual variations in metabolism of endo- and exogenous compounds due to genetic influences, leading to enhanced side effects or toxicity of drugs (for example, poor versus fast metabolisers) or to different clinical effects (metabolism of steroid hormones). See **ecogenetics, pharmacogenetics, toxicogenetics**

- **polyphagia**
excessive eating or hunger, a characteristic of uncontrolled diabetes mellitus.

- **polyuria**
the passage of a large amount of urine in a given period, a characteristic of diabetes mellitus.

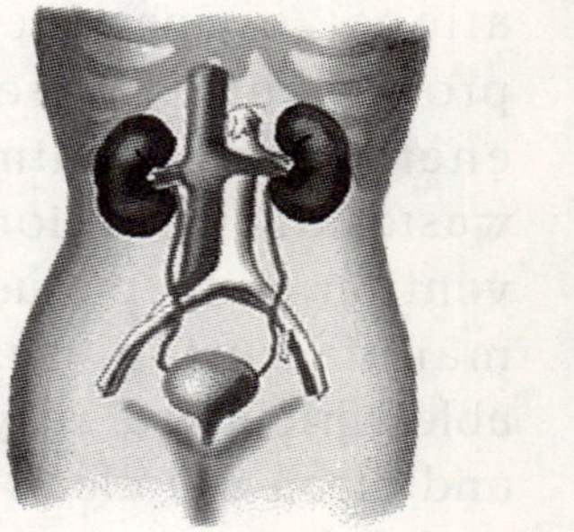

- **population**
1. in statistics, the totality of items under consideration. A clearly defined part of a population is called a subpopulation. In the case of a random variable, the probability distribution is considered as defining the population of that variable. The term 'population segment' is sometimes used as a synonym for subpopulation.
2. a group or number of people living within a speci-

fied area or sharing similar characteristics (such as occupation or age).

- **population at risk**
  1. number of persons who can and may develop an adverse health effect and who are potentially exposed to a risk factor under study: for example, all people in a population who have not developed immunity to an infectious disease are at risk of developing that disease if they are exposed to it. People already having chronic disease are excluded from the population at risk in studies of the incidence of the disease.
  2. a population subgroup that is more likely to be exposed to a chemical, or is more sensitive to a chemical, than is the general population.

- **population critical concentration (PCC)**
  concentration of a substance in the critical organ at which a specified percentage of the exposed population has reached the individual critical organ concentration. The percentage indicated by PCC-10 for 10 %, PCC-50 for 50 % etc. (similar to the use of the term LD_{50}).

- **population effect**
  absolute number or incidence rate of cases occurring in a group of people.

- **pore water**
  water occupying space between sediment or soil particles (syn. Interstitial Water).

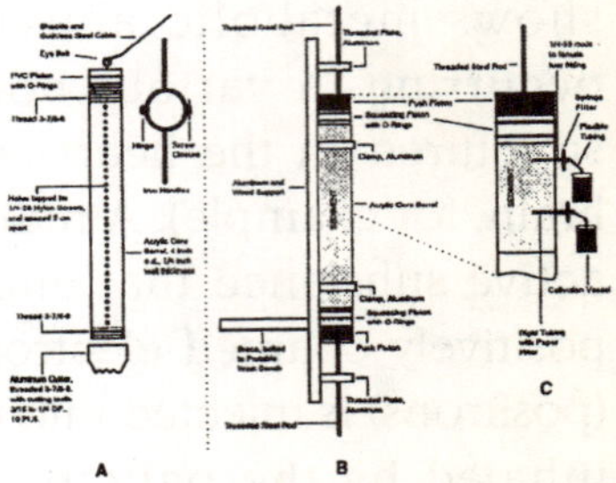

- **pore**
  a very small open space in a rock or granular material. Also see **interstice**.

- **porosity**
  the capacity of soil of rock to hold water.

- **porphyria**
  disturbance of porphyrin metabolism characterised by increased formation, accumulation, and excretion of porphyrins and their precursors.

• **positive displacement pump**
a type of piston, diaphragm, gear or screw pump that delivers a constant volume with each stroke. Positive displacement pumps are used as chemical solution feeders.

• **positron emission tomography (PET)**
a computerised radiographic technique that shows metabolic activity occurring in various body structures (in the heart and brain, for example). A radioactive substance that emits positively charged electrons (positrons) is injected into or inhaled by the patient, in whom they combine with negatively charged electrons. The resulting gamma radiation is converted into colour-coded images.

• **posology**
study of dose in relation to the physiological factors that may influence response such as age of the exposed organisms.

• **postchlorination**
the addition of chlorine to the plant effluent, following plant treatment, for disinfection purposes.

• **postpartum thyroiditis**
an autoimmune condition resulting in transient thyrotoxicosis followed by hypothyroidism occurring after parturition (giving birth). It may recur with subsequent pregnancies.

• **potable water**
raw or treated water that is considered safe to drink.

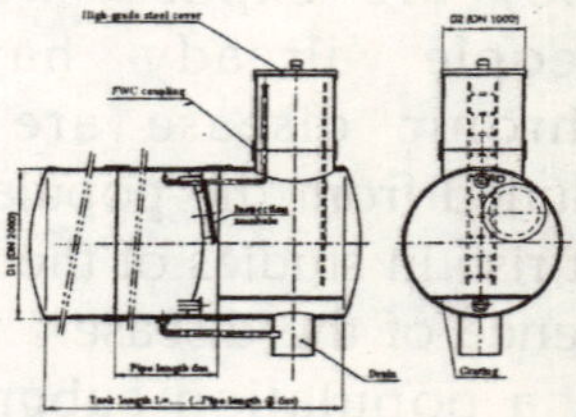

• **potassium (K)**
an essential element found primarily in the cells of the body. It helps to maintain proper muscle and nerve function. It also helps in some important metabolic processes. Blood serum potassium levels usually are determined to help evaluate heart rhythm irregularities, neuromuscular disorders and kidney function.

- **potency**
  expression of chemical or medicinal activity of a substance as compared to a given or implied standard or reference.

- **potentially responsible party (PRP)**
  a company, government, or person legally responsible for cleaning up the pollution at a hazardous waste site.

- **potentiation**
  1. dependent action in which a substance or physical agent at a concentration or dose that does not itself have an adverse effect enhances the harm done by another substance or physical agent. See **additive effect, antagonism, synergism.**
  2. potentiation is the ability of one chemical to enhance the activity of another chemical to an extent greater than the simple summation of the two expected activities.

- **potentiometric surface**
  the level to which water will rise in cased wells or other cased excavations into aquifers, measured as feet above mean sea level.

- **power**
  the probability of correctly detecting an actual toxic effect .

- **power factor**
  the ratio of the true power passing through an electric circuit to the product of the voltage and amperage in the circuit. This is a measure of the lag or load of the current with respect to the voltage.

- **practical certainty**
  the concept of practical certainty involves the determination of a numerically specified low risk or socially acceptable risk which may be used in decision making where absolute certainty is not possible.

- **prechlorination**
  the addition of chlorine at the headworks of the plant prior to other treatment processes mainly for disinfection and control of taste, odour and aquatic growth. Also applied to aid in coagulation and settling.

- **precipitate**

  1. an insoluble, finely divided substance which is a product of a chemical reaction within a liquid.
  2. the separation from solution of an insoluble substance.
  3. removal of solids from liquid waste so that the hazardous solid portion can be disposed of safely, removal of particles from airborne emissions.
  4. the process by which atmospheric moisture falls onto land or water surface as rain, snow, hail, or other forms of moisture. .

- **precision**

  1. a measure of the reproducibility within a data set. Precision can be measured both within a laboratory and between laboratories using the same test method and toxicant.
  2. the ability of an instrument to measure a process variable and to repeatedly obtain the same result. The ability of an instrument to reproduce the same results.

- **pre-cordial**

  pertaining to the region over the heart and lower thorax.

- **precursor**

  substance from which another, usually more biologically active, substance is formed.

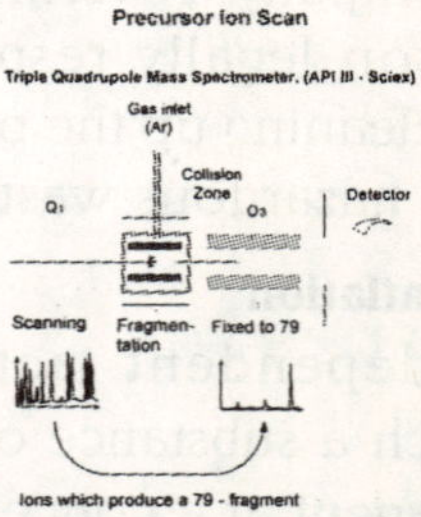

- **precursor, THM (pre-curse-or)**

  natural organic compounds found in all surfaces and groundwater. These compounds MAY react with halogens (such as chlorine) to form trihalomethanes (try-HAL-o-METH-hanes) (THMs). They must be present in order for THMs to form.

- **predicted environmental concentration**

  this is the estimated concentration of a chemical in an environmental compartment calculated from available information on its properties, its use and discharge patterns, and the quantities involved.

- **predictive validity**
  reliability of a measurement expressed in terms of its ability to predict the criterion: an example would be an academic aptitude test that was validated against subsequent academic performance.

- **predictive value**
  percentage of positive results that are true positives or of negative results that are true negatives. See **sensitivity, specificity**

- **preneoplastic**
  before the formation of a tumour.

- **prescriptive**
  water rights which are acquired by diverting water and putting it to use in accordance with specified procedures. These procedures include filing a request to use unused water in a strewn, river or lake with a state agency.

- **pressure control**
  a switch which operates on changes in pressure. Usually this is a diaphragm pressing against a spring. When the force on the diaphragm overcomes the spring pressure, the switch is activated.

- **pressure head**
  the vertical distance (in feet) equal to the pressure (in psi) at a specific point. The pressure head is equal to the pressure in psi times 2.31 ft/psi.

- **pre-treatment**
  methods used by industry and other non-household sources of wastewater to remove, reduce, or alter the pollutants in wastewater before discharge to a STW.

- **prevalence**
  the number of existing disease cases in a defined population during a specific time period [contrast with incidence].

- **prevalence rate (ratio)**
  total number of individuals who have an attribute or disease at a particular time (or during a particular period) divided by the population at risk of having the attribute or disease at this point in time or midway

through the period. See **population at risk**.

- **prevalence study**
  an epidemiological study which examines the relationships between diseases and exposures as they exist in a defined population at a particular point in time.

- **prevalence survey**
  the measure of the current level of disease(s) or symptoms and exposures through a questionnaire that collects self-reported information from a defined population.

- **prevention**
  actions that reduce exposure or other risks, keep people from getting sick, or keep disease from getting worse.

- **primacy**
  the responsibility for ensuring that a law is implemented, and the authority to enforce a law and related regulations. A primacy agency has the primary responsibility for administrating and enforcing regulations.

- **primary element**
  an instrument which measures (senses) a physical condition or variable of interest. Floats and thermocouples are examples of primary elements. Also called a sensor.

- **primary pollutant**
  see **pollutant.**

- **primary protection standard**
  accepted maximum level of a pollutant (or its indicator) in the target organism, or some part thereof, or an accepted maximum intake of a pollutant or nuisance into the target under specified circumstances.

- **primary treatment**
  first stage of wastewater treatment in which solids are removed by screening and settling.

- **primary wavelength**
  the wavelength at which a chemical has the greatest absorbance.

- **prime**
  the action of filling a pump casing with water to remove the air. Most pumps must be primed before start

up or they will not pump any water.

- **prior appropriation**
a doctrine of water law that allocates the right to use water on a first-come first-serve basis.

- **probit**
probability unit obtained by adding 5 to the normal deviates of a standardized normal distribution of results from a dose response study: addition of 5 removes the complication of handling negative values. A plot of probit against the logarithm of dose or concentration gives a linear plot if the distribution of response is a logarithmic normal one. Estimates of the LD_{50} and ED_{50} (or LC_{50} and EC_{50}) can be obtained from this plot.

- **procarcinogen**
substance that has to be metabolised before it can induce malignant tumours.

- **process variable**
a physical or chemical quantity which is usually measured and controlled in the operation of a water treatment plant or an industrial plant.

- **process waste-water**
any water which comes into contact with any raw material, product, by-product, or waste.

- **product water**
water that has passed through a water treatment plant. All the treatment processes are completed or finished. This water is the product from the water treatment plant and is ready to be delivered to the consumers. Also called finished water.

- **proficiency test sample (PT)**
a sample, the composition of which is unknown to the analyst, and is provided to test whether the analyst/laboratory can produce analytical results within specified acceptance criteria.

- **profile**
a drawing showing elevation plotted against distance, such as the vertical section or side view of a pipeline.

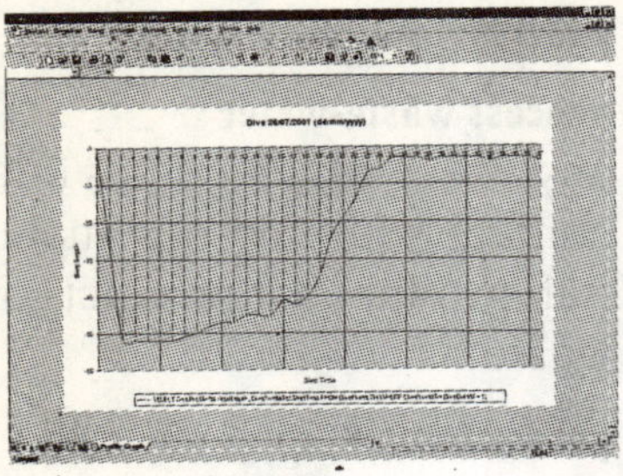

- **prokaryote**
unicellular organism, characterised by the absence of a membrane-enclosed nucleus. Prokaryotes include bacteria, blue-green algae and mycoplasmas. See **eukaryote**.

- **promoter (carcinogenicity)**
this is an agent which increases tumour production by another substance when applied to susceptible organisms after their exposure to the first substance.

- **promoter (in oncology)**
agent that induces cancer when administered to an animal or human being who has been exposed to a cancer initiator. See **initiator**.

- **prophage**
latent state of a phage genome in a lysogenic bacterium.

- **proportional mortality rate (ratio) (PMR)**
number of deaths from a given cause in a specified time period, per 100 or per 1000 total deaths in the same time period: can give rise to misleading conclusions if used to compare mortality experience of populations with different causes of death.

- **prospective cohort study**
see **cohort study**.

- **prospective study**
an epidemiological study which examines the development of disease in a group of persons determined to be presently free of the disease.

- **prostate specific antigen test**
a blood test which measures a patient's level of prostate specific antigen. If elevated, the antigens may be an indicator of prostate cancer.

- **prostate specific antigen(PSA)**
a substance manufactured mainly by the prostate.

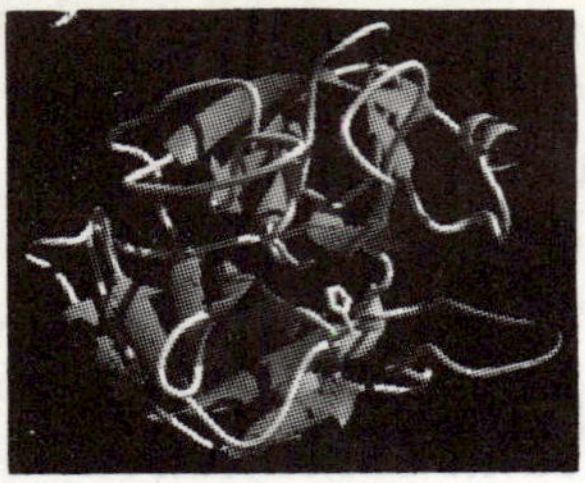

- **prostatectomy**
surgical removal of the prostate.

- **prostatic acid; phosphatase test**
a blood substance that rises above normal in many patients when prostate cancer has spread beyond the prostate.

- **protein binding**
this is the property of having a physico-chemical affinity for protein.

- **proteinuria**
excretion of excessive amounts of protein (derived from blood plasma or kidney tubules) in the urine.

- **prothrombin**
a plasma protein, one of the coagulation factors and the precursor of thrombin, prothrombin is synthesised in the liver with the help of vitamin K. The prothrombin time (PT) test measures blood clotting time, evaluates bleeding disorders and liver damage, and monitors the use of anticoagulant drugs.

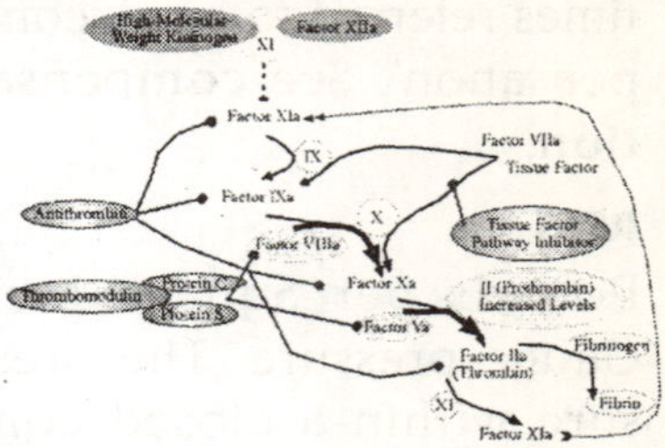

- **protocol**
the detailed plan for conducting a scientific procedure. A protocol for measuring a chemical in soil, water or air describes the way in which samples should be collected and analysed.

- **prussian blue**
a blue paste or liquid (often on a paper like carbon paper) used to show a contact area. Used to determine if gate valve seats fit properly.

- **pseudo-adaptation**
apparent adaptation of an organism to changing conditions of the environment (especially chemical) associated with stresses in biochemical systems that exceed the limits of normal (homeostatic) mechanisms. Essentially there is a tempo-

rary concealed pathology that later on can be manifested in the form of explicit pathological changes sometimes referred to as 'decompensation'. See **compensation.**

- **PSIG**
  Pounds per Square Inch Gauge pressure. The pressure within a closed container or pipe measured with a gauge in pounds per square inch. See **gauge pressure**.
- **psychosis**
  any major mental disorder characterised by derangement of the personality and loss of contact with reality.
- **psychotropic**
  exerting an effect upon the mind, capable of modifying mental activity.
- **public health action**
  a list of steps to protect public health.
- **public health assessment(PHA)**
  a document that examines hazardous substances, health outcomes, and community concerns at a hazardous waste site to determine whether people could be harmed from coming into contact with those substances. The PHA also lists actions that need to be taken to protect public health [compare with **health consultation**].
- **public health hazard**
  sites that pose a public health hazard because of long-term exposures (greater than 1 year) to sufficiently high levels of hazardous substances or radio-nuclides that could result in harmful health effects.
- **public health hazard categories**
  public health hazard categories are statements about whether people could be harmed by conditions present at the site in the past, present, or future. One or more hazard categories might be appropriate for each site. The five public health hazard categories are no public health hazard, no apparent public health hazard, indeterminate public health hazard, public health hazard, and urgent public health hazard.

- **public health impact assessment**
applying risk assessment to a specific target population of known size, giving as the end product a quantitative statement about the number of people likely to be affected in a particular population. See **risk assessment.**

- **public health statement**
the public health statement explains how people might be exposed to a specific substance and describes the known health effects of that substance.

- **public health surveillance**
the ongoing, systematic collection, analysis, and interpretation of health data. This activity also involves timely dissemination of the data and use for public health programmes.

- **public meeting**
a public forum with community members for communication about a site.

- **public water system**
any water system that regularly supplies piped water to the public for consumption.

- **public water system**
a system for the provision to the public of piped water for human consumption, If such system has at least fifteen service connections or regularly serves an average of at least 25 individuals at least 60 days out of the year. The term includes: 1) any collection, treatment, storage, and distribution facilities under control of the operator of such system and used primarily in connection with such system, and 2) any collection or pre-treatment storage facilities not under such control which are used primarily in connection with such system. A public water system is either a 'community water system' or a 'non-community water system'.

- **pulmonary**
pertaining to the lungs.

- **pulmonary alveoli**
  the pulmonary alveoli are minute air-filled sacs in the vertebrate lung, thin walled and surrounded by blood vessels.

- **pulmonary compartment**
  portion of the respiratory tract in which gas exchange occurs (including alveoli and respiratory bronchioles)

- **pumping station**
  mechanical devices installed in sewer or water systems or other liquid-carrying pipelines that move the liquids to a higher level.

- **pumping water level**
  the vertical distance in feet from the centreline of the

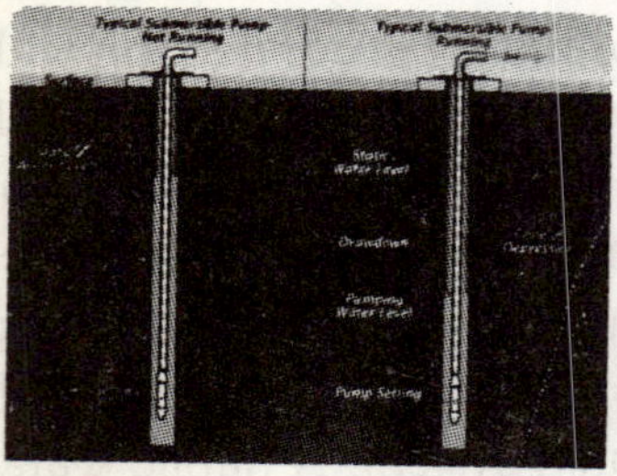

  pump discharge to the level of the free pool while water is being drawn from the pool.

- **purgative**
  see **cathartic**, **laxative**.

- **purveyor, water**
  an agency or person that supplies water (usually potable water).

- **putrefaction**
  biological decomposition of organic matter, with the production of ill-smelling and tasting products, associated with anaerobic (no oxygen present) conditions.

- **pyrexia**
  condition in which the temperature of a human being or mammal is above normal.

- **pyrogen**
  any substance that produces fever.

- **qualitative**
  1. a test in which the final result is positive/negative, reactive/non-reactive, present/absent.
  2. descriptive of kind, type or direction, as opposed to size, magnitude or degree.

- **quality assurance**
  all those planned and systematic actions necessary to provide adequate confidence that a product or service will satisfy given requirements for quality. See

good laboratory practice, quality control.

- **quality assurance and quality control (QA/QC)**
  a system of procedures, checks and audits to judge and control the quality of measurements and reduce the uncertainty of data. Some quality control procedures include having more than one person review the findings and analyse a sample at different times or laboratories to see if the findings are similar.

- **quality control**
  the overall system of technical activities whose purpose is to measure and control the quality of a product or service so that it meets the needs of users.

- **quantal effect**
  condition that can be expressed only as 'occurring' or 'not occurring', such as death or occurrence of a tumour. See **graded effect, stochastic effect, all-or-none effect**.

- **quantitative**
  1. a test in which the final result is expressed as a specific quantity — ng, mg, mg.
  2. descriptive of size, magnitude or degree.

- **quantitative structure-activity relationship (QSAR)**
  quantitative association between the physico-chemical properties of a substance and/or the properties of its molecular substructures and its biological properties including its toxicity. See **surrogate, structure-activity relationship**.

- **quicklime**
  a material that is mostly calcium oxide (CaO) or calcium oxide in natural association with a lesser amount of magnesium oxide. Quicklime is capable of combining with water to form hydrated lime. Also see **hydrated lime**.

• **radial to impeller**
perpendicular to the impeller shaft. Material being pumped flows at a right angle to the impeller.

• **radical**
a group of atoms that is capable of remaining unchanged during a series of chemical reactions. Such combinations (radicals) exist in the molecules of many organic compounds. Sulphate $(SO_4)^{2-}$ is an inorganic radical.

• **radio-active**
emitting radiation that results from atomic disintegration.

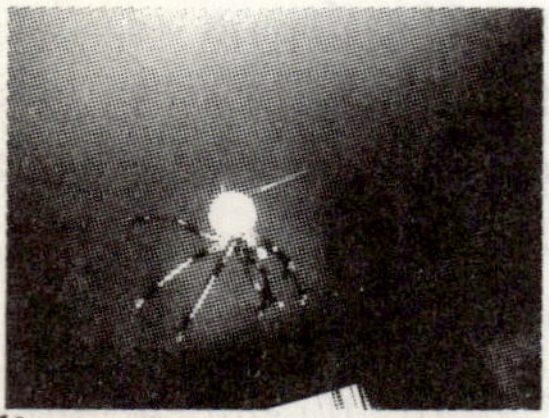

• **radio-active energy attenuation**
reduction of intensity of an X-ray beam by the partial absorption of radiation, as by a radio-lucent structure. Soft tissue is radio-lucent to some extent. In attenuation, the number of X-rays left in an X-ray beam after penetration through matter is reduced, but never reaches zero.

• **radio-active iodine excretion test**
a test of thyroid function in which the patient is given radioactive iodine by mouth. The amount of radioactivity excreted in the urine and/or accumulated in the thyroid reflects thyroid activity.

• **radio-active waste**
any waste that emits energy as rays, waves, or streams of

energetic particles. Radioactive materials are often mixed with hazardous waste, usually from nuclear reactors, research institutions, or hospitals.

- **radioimmunoassay (RIA)**
  see **assay**.

- **radioisotopes**
  elements in a form with unstable nuclei that achieve a more stable nuclear configuration by emitting excess energy as radiation (radioactive decay). Radioisotopes can be used to label certain chemicals to measure body function.

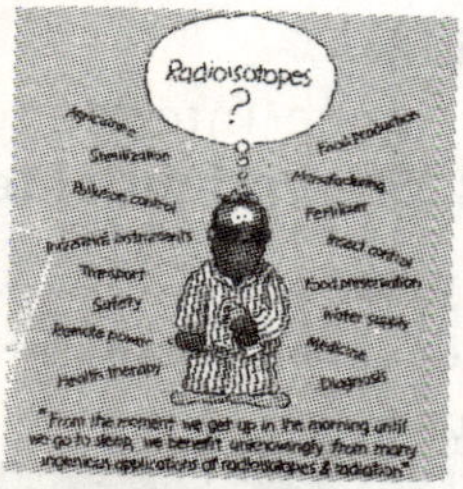

- **radionuclide**
  any radioactive isotope (form) of any element.

- **radiopaque dye**
  a substance or chemical (such as certain iodine compounds) that is given before an X-ray procedure is performed to outline the interior of hollow organs. It does not permit the passage of X-rays.

- **radio-pharmaceuticals**
  drugs that spontaneously emit nuclear particles. Different radio-pharmaceuticals are chosen according to the organ and purpose (diagnostic, therapeutic, or research). The radioactive material allows differentiation between healthy and diseased tissue.

- **radon**
  a colourless, naturally occurring gas formed by radioactive decay of radium atoms. Radon accumulating in basements and other areas of buildings without proper ventilation has been identified as a leading cause of lung cancer.

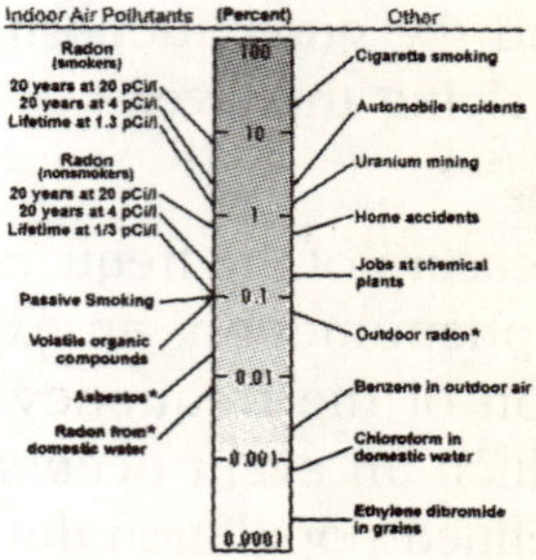

- **râles**
  see **crepitations**.

- **random sample**
  subset of a population that is arrived at by selecting units such that each possible unit has a fixed and deter-

minate probability of selection. See **biased sample, sample**.

• **range**
the spread from minimum to maximum values that an instrument is designed to measure. Also see **span and effective range**.

• **ranney collector**
this water collector is constructed as a dug well from 12 to 16 feet (3.5 to 5 m) in diameter that has been sunk as a caisson near the bank of a river or lake. Screens are driven radially and approximately horizontally from this well into the sand and the gravel deposits underlying the river.

• **rate**
measure of the frequency of a phenomenon: an expression of the frequency with which an event occurs in a defined population during a specified time interval.

• **rate difference (RD)**
absolute difference between two rates, for example, the difference in incidence rate between a population group exposed to a causal factor and a population group not exposed to the factor: in comparisons of exposed and unexposed groups, the term 'excess rate' may be used as a synonym for rate difference.

• **rate ratio (RR)**
in epidemiology, the value obtained by dividing the rate in an exposed population by the rate in an unexposed population.

• **ratticide**
substance intended to kill rats. See **rodenticide.**

• **raw water**
water prior to any treatment or use.

• **RD50**
the concentration of a test article which will depress the respiration rate of mice by 50%.

• **reactivity**
refers to those hazardous wastes that are normally

unstable and readily undergo violent chemical change but do not explode.

- **readily biodegradable**
  arbitrary classification of substances that have passed certain specified screening tests for ultimate biodegradability. These tests are so stringent that such compounds will be rapidly and completely biodegraded in a wide variety of aerobic environments.

- **re-aeration**
  the introduction of air through forced air diffusers into the lower layers of the reservoir. As the air bubbles form and rise through the water, oxygen from the air dissolves into the water and replenishes the dissolved oxygen. The rising bubbles also cause the lower waters to rise to the surface where oxygen from the atmosphere is transferred to the water. This is sometimes called surface re-aeration.

- **reagent**
  a substance used to produce a specific chemical reaction. It may be used to detect, measure, or prepare other substances.

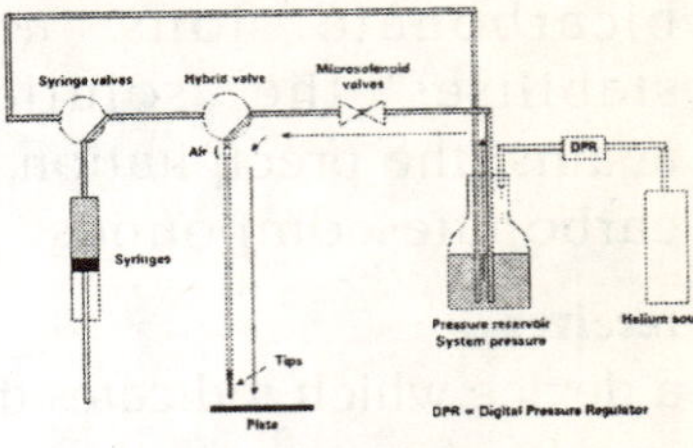

ASAP System - Dispense Reagent

- **reasonable maximum exposure (RME)**
  highest exposure that is reasonably expected to occur. Typically the 95% upper confidence limit of the toxicant distribution is used. If only a few data points (6-10) are available, the maximum detected concentration is used.

- **recalcitrance**
  ability of a substance to remain in a particular environment in an unchanged form.
  See **persistence**.

- **re-carbonation**
  a process in which carbon dioxide is bubbled into the water being treated to lower the pH. The pH may also be lowered by the addition of acid. Re-carbonation is the final stage in the lime-soda ash soften-

ing process. This process converts carbonate ions to bicarbonate ions and stabilises the solution against the precipitation of carbonate compounds.

- **receiver**
a device which indicates the result of a measurement. Most receivers in the water utility field use either a fixed scale and movable indicator (pointer) such as pressure gauge or a movable scale and movable indicator like those used on a circular-flow recording chart. Also called an indicator.

- **receiving waters**
a river, lake, ocean, stream, or other body of water into which wastewater or treated effluent is discharged.

- **receptor**
high affinity binding site for a particular toxicant. See **target, target organ**.

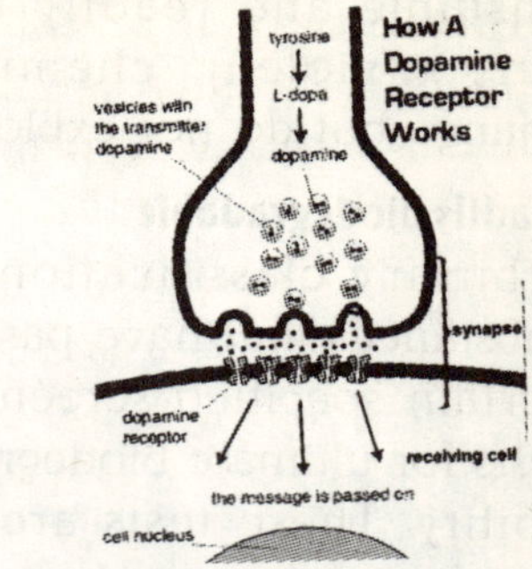

- **receptor population**
people who could come into contact with hazardous substances (see **exposure pathway**).

- **recharge area**
generally, an area that is connected with the underground aquifer(s) by a highly porous soil or rock layer. Water entering a recharge area may travel for miles underground.

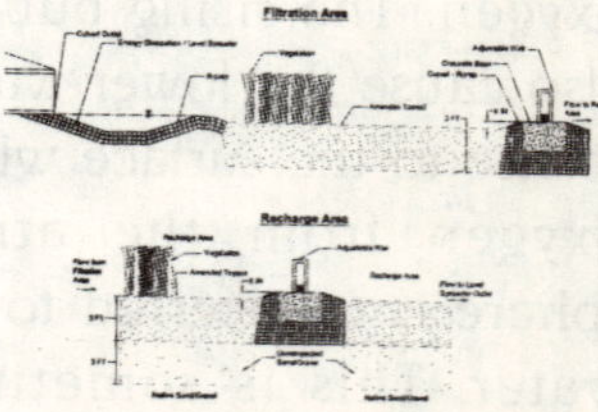

Schematic Recharge Area Profile

- **recharge rate**
the quantity of water per unit time that replenishes or refills an aquifer.

- **recharge**
process by which rain water (precipitation) seeps into the ground-water system.

- **recommended limit**
this regulatory value is the maximum concentration of a potentially toxic substance which is believed to be safe. Such limits often have no legal backing in which case a control or statutory guide level which should not be exceeded under any circumstances may be set. See **control limit**.

- **recorder**
a device that creates a permanent record, on a paper chart or magnetic tape, of the changes of some measured variable.

- **recovery**
process leading to partial or complete restoration of a cell, tissue, organ or organism following its damage from exposure to a harmful substance or agent.
Term used in analytical and preparative chemistry to denote the fraction of the total quantity of a substance recoverable following a chemical procedure. See **recovery factor.**

- **recovery factor**
fraction or percentage of the total quantity of a substance extracted under specified conditions.

- **rectum**
the last five or six inches of the intestine leading to the outside of the body.

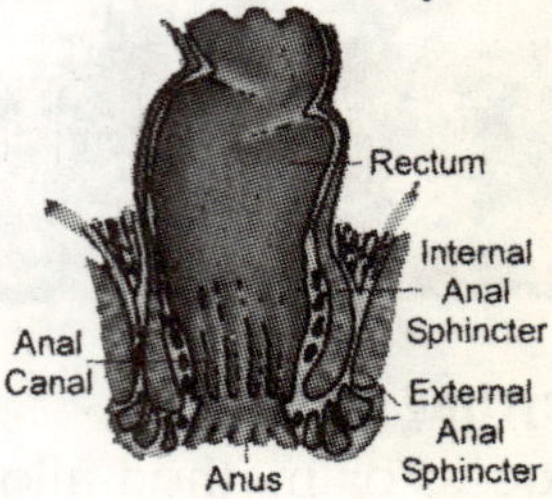

- **recycled water**
process wastewater or treatment facility effluent which is re-circulated to the same process.

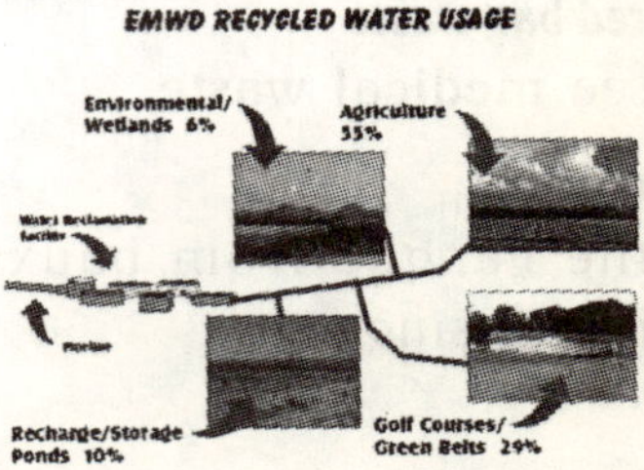

- **recycle-reuse**
 the process of minimising the generation of waste by recovering usable products that might otherwise become waste. Examples are the recycling of aluminium cans, wastepaper, and bottles.

- **recycling**
 reusing materials and objects in original or changed forms rather than discarding them as wastes.

- **recycling (of waste)**
 process or method allowing for the recovery of some value from a waste, either as re-usable material or as energy.

- **red bag waste**
 see **medical waste.**

- **red mud**
 the gangue from bauxite processing.

- **red tide**
 a proliferation of ocean plankton that may kill large numbers of fish. This natural phenomenon may be stimulated by the addition of nutrients.

- **redox**
 a term used to abbreviate a reduction-oxidation reaction.

- **redox potential**
 reduction-oxidation potential measured against a standard electrode.

- **reducing agent**
 1. in a reduction reaction (which always occurs simultaneously with an oxidation reaction), the reducing agent is the chemical or substance that (1) combines with oxygen or(2) loses electrons to the reaction (see **oxidation, oxidising agent**). 2. any substance, such as a base metal (iron) or the sulphide ion (S^{2-}), that will

readily donate (give up) electrons. The opposite is an oxidising agent.

- **reduction**
reduction is the addition of hydrogen, removal of oxygen, or the addition of electrons to an element or compound. Under anaerobic conditions (no dissolved oxygen present), sulphur compounds are reduced to odour-producing hydrogen sulphide (H_2S) and other compounds. The opposite of oxidation.

- **reference concentration**
term used for an estimate of air exposure concentration to the human population (including sensitive subgroups) that is likely to be without appreciable risk of deleterious effects during a lifetime. See **acceptable daily intake, dose.**

- **reference distribution**
statistical distribution of reference values.

- **reference dose**
term used for an estimate (with uncertainty spanning perhaps an order of magnitude) of a daily exposure to the human population (including sensitive subgroups) that is likely to be without appreciable risk of deleterious effects during a lifetime. See **acceptable daily intake, dose.**

- **reference group**
see **reference sample group**.

- **reference individual**
person selected with the use of defined criteria for comparative purposes in a clinical study.

- **reference interval**
area between and including two reference limits, for example the percentiles 2.5 and 97.5.

- **reference limit**
boundary value defined so that a stated fraction of the reference values is less than or exceeds that boundary value with a stated probability.

- **reference material**
substance for which one or more properties are sufficiently well established to be used for the calibration of an apparatus, the assessment of a measurement method, or for assigning

values to other substances. See **calibration material**, **standard material**.

- **reference population**
group of all reference individuals used to establish criteria against which a population that is being studied can be compared.

- **reference sample group**
selected reference individuals, statistically adequate numerically to represent the reference population.

- **reference toxicant**
the toxicant used in performing toxicity tests to indicate the sensitivity of a test organism and to demonstrate the laboratory's ability to perform the test correctly and obtain consistent result.

- **reference value**
measured value of a property in a reference individual or sample from a reference individual.

- **reference**
a physical or chemical quantity whose value is known exactly, and thus is used to calibrate or standardise instruments.

- **refine**
to remove impurities.

- **registry**
a systematic collection of information on persons exposed to a specific substance or having specific diseases [see **exposure registry and disease registry**].

- **regulatory dose**
the expected dose resulting from human exposure to a substance at the level at which it is regulated in the environment.

- **relative odds**
see **odds ratio.**

- **relative risk**
this term may have any of the following meanings depending upon context.
1. ratio of the risk of disease or death among the exposed to the risk among the unexposed: this usage is synonymous with 'risk ratio'.
2. alternately, the ratio of the cumulative incidence rate in the exposed to the cumulative incidence rate in the unexposed, i.e. the cumulative incidence ratio.
3. sometimes used as a synonym for 'odds ratio'. See **risk**.

- **relaxation time**
time for a particle to reach 1/*e* of its final velocity from an initial velocity or from rest when subjected to an external force. It's an indicator of a particle's ability to adjust to changes in flow velocity.

- **release**
any spilling, leaking, pumping, pouring, emitting, emptying, discharging, injecting, escaping, leaching, dumping, or disposing into the environment of a hazardous or toxic chemical, or extremely hazardous substance.

- **reliquifaction (re-lick-we-fack-shun)**
the unit of dose equivalent from ionising radiation to the total body or any internal organ or organ system. A millirem (mrem) is 1/1000 of a rem.

- **remedial action**
the actual construction or clean-up phase of a site cleanup.

- **remedial investigation (RI)**
an in-depth study (including sampling of air, soil, water and waste) of a contaminated site needing remediation to determine the nature and extent of contamination. The remedial investigation (RI) is usually combined with a feasibility study (FS).

- **remediation**
correction or improvement of a problem, such as work that is done to clean up or stop the release of chemicals from a contaminated site. After investigation of a site, remedial work may include removing soil and/or drums, capping the site or collecting and treating the contaminated fluids.

- **renal**
pertaining to the kidneys.

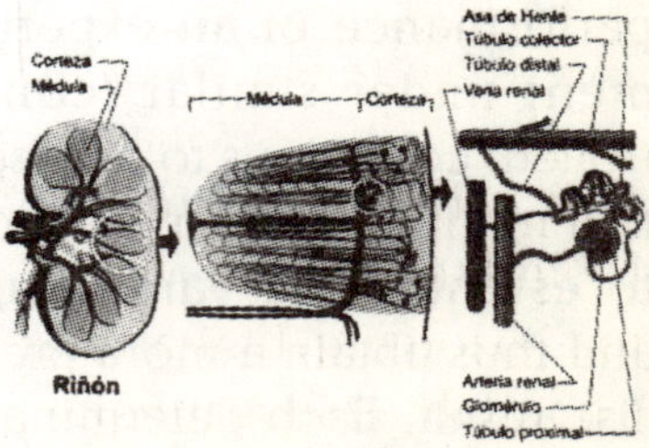

• **repeat compliance period**
any subsequent compliance period after the initial compliance period.

• **repellent**
substance used mainly to repel blood sucking insects in order to protect man and animals: also used to repel mammals, birds, rodents, mites, plant pests, etc.

• **replicate**
each of several experimental units that are tested simultaneously using the same experimental conditions

• **replicate sampling**
act of taking multiple samples concurrently under comparable conditions; may be accomplished by taking samples adjacent in time or space.

• **replication**
1. duplicated or repeated performance of an experiment under similar (controlled) conditions to reduce to a minimum the error, and to estimate the variations and thus obtain a more precise result. Each determination, including the first is called a replicate.

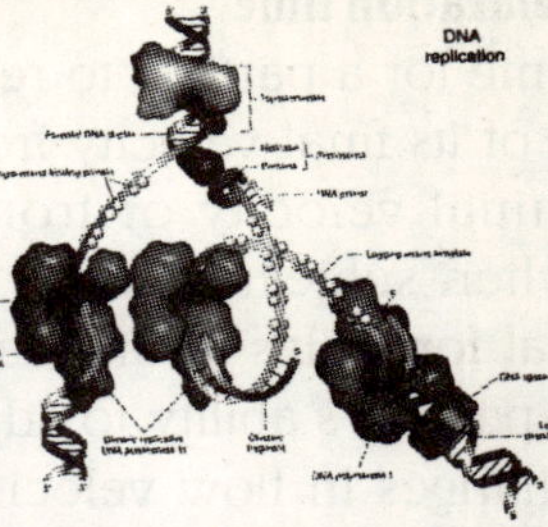

2. process whereby the genetic material is duplicated.

• **representative sample**
a portion of material or water that is as nearly identical in content and consistency as possible to that in the larger body of material or water being sampled.

• **reproducibility**
closeness of agreement between test results obtained under reproducibility conditions. See **reproducibility conditions**.

• **reproducibility conditions**
situation where test results are obtained with the same method on identical test material in different laboratories with different operators using different equipment.

• **reproductive toxicant**
substance or preparation that produces non-heritable harmful effects on the

progeny and/or an impairment of male and female reproductive function or capacity. See **teratogen.**

- **reproductive toxicology**
  study of the adverse effects of substances on the embryo, foetus, neonate and pre-pubertal mammal and the adult reproductive and neuro-endocrine systems. See **embryo, foetus, neonate**.

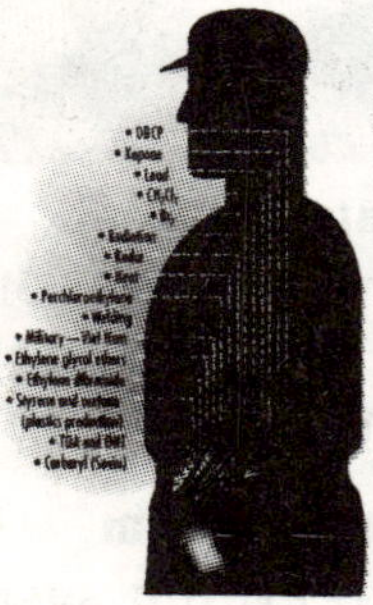

- **reserve air volume**
  air retained in the deepest part of the lung where there is insignificant convection.

- **reserve capacity**
  physiological or biochemical capacity that may be available to maintain homeostasis when the body or an organism is exposed to an environmental change.

- **reservoir**
  any natural or artificial holding area used to store, regulate, or control water.

- **resident species**
  a species that is regularly present at a specified site for some portion of its life span .

- **residual chlorine**
  the amount of free and/or available chlorine remaining after a given contact time under specified conditions.

- **residual disinfectant concentration ('c' in ct calculations)**
  the concentration of disinfectant measured in mg/l in a representative sample of water.

- **residue**
  the dry solids remaining after the evaporation of a sample of water or sludge. Also see total dissolved solids

- **resistance (in toxicology)**
  ability to withstand the effect of various factors including potentially toxic substances.

- **resorptive effect**
  action of a substance after its resorption from the gut into the blood.

- **resource recovery**
  the extraction of useful materials or energy from solid waste. Such materials can include paper, glass, and metals that can be reprocessed for re-use. Resource recovery also is employed in pollution prevention.

- **respirable fraction**
  fraction of aerosol that can reach the gas exchange region of the human respiratory system.

- **respirable particles**
  those materials that are deposited in the gas-exchange region of the lung.

- **respiration**
  the process in which an organism uses oxygen for its life processes and gives off carbon dioxide.

- **respirator**
  any of a variety of devices that limit the inhalation of toxic materials. They range from disposable dust masks to self-contained breathing apparatus. All have specific uses and limitations.

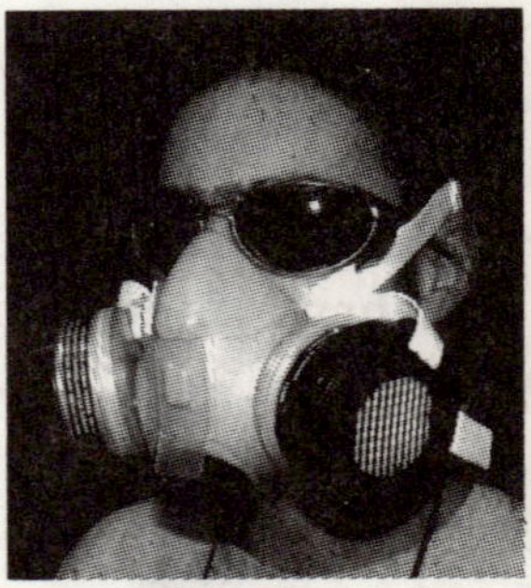

- **respiratory rate**
  the average rate of breathing; e.g., breaths per minute.

- **respiratory system**
  the breathing system, including the lungs and air passages (trachea or windpipe, larynx, mouth, and nose), as well as the associated system of nerves and circulatory supply.

- **respiratory tract**
  composed of the conducting airway including the nose, mouth, pharynx, larynx, trachea, bronchi, bronchioles, and alveoli.

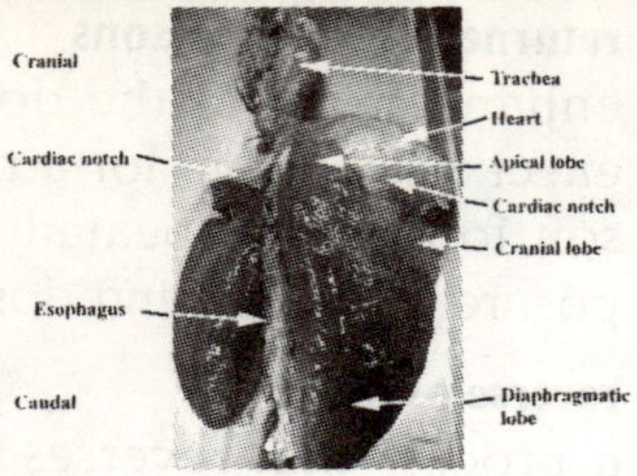

- **response**

  1. proportion of an exposed population with a defined effect or the proportion of a group of individuals that demonstrate a defined effect in a given time at a given dose rate. See **dose-response relationship**.

  2. reaction of an organism or part of an organism (such as a muscle) to a stimulus.

- **retention**

  1. holding back within the body or within an organ, tissue or cell of matter that is normally eliminated. See **elimination**.

  2. holding in memory what has been learned, for later use as recall, recognition or relearning.

  3. amount of a substance that is left from the total absorbed after a certain time following exposure: if the retention follows a course in relation to time that is a first order process, it may be described in terms of biological half-life (half-time). See **half-life**.

- **reticulocyte**

  a young red blood cell containing a network of basophilic substances.

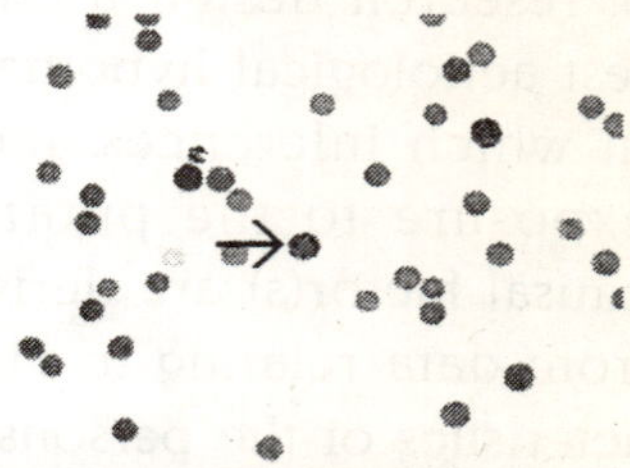

- **reticuloendothelial system**

  this is the system of cells which have the ability to take up and retain certain dyes and particles ingested into a living animal. This term has generally been replaced by the term, mononuclear phagocyte system.

- **retinopathy**

  this condition of the eye is associated with diabetes mellitus and is progressively characterised by microaneurysms, intraretinal punctuate haemorrhages, yellow waxy exudates, cotton-wool patches, and macular oedema.

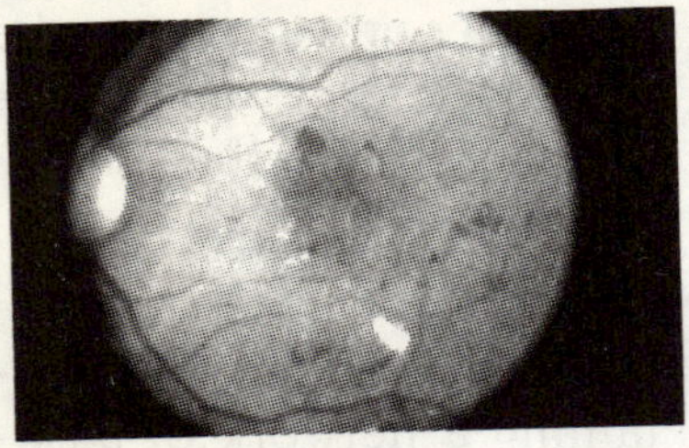

- **retrospective study**

  1. research design used to test aetiological hypotheses in which inferences about exposure to the putative causal factor(s) are derived from data relating to characteristics of the persons or organisms under study or to events or experiences in their past. The essential feature is that some of the persons under study have the disease and their characteristics and past experiences are compared with those of other, unaffected persons. Persons who differ in the severity of the disease may also be compared. See **case control study**.

  2. an epidemiological study which compares diseased persons with non-diseased persons and works back in time to determine exposures.

- **returned effect of poisons**

  enhancement of the dose-effect relationship for a poison following repeated exposure to decreasing doses.

- **reverse osmosis**

  a process that reverses (by the application of pressure)

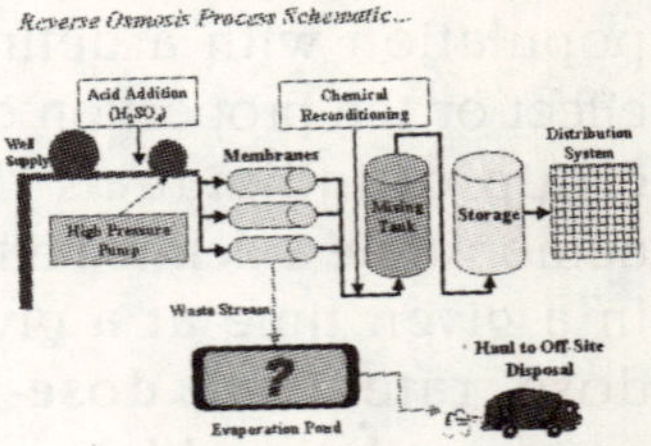

  the flow of water in the natural process of osmosis so that it passes from the more concentrated to the more dilute solution.

- **reverse transcription**

  process by which an RNA molecule is used as a template to make a single-stranded DNA copy.

- **reversible alteration**

  change from normal structure or function, induced by a substance or other agent(s), that returns to normal status or within normal limits after cessation of exposure.

- **reversible effect**
an effect which is not permanent, especially adverse effects which diminish when exposure to a toxic chemical is ceased.

- **Reynold's number**
flow similitude parameter, expressed as the ratio of the inertial force of the gas to the friction force of the gas moving over the surface of an object.

- **rfd (reference dose)**
the daily exposure level which, during a lifetime of a human, appears to be without appreciable risk on the basis of all facts known at the time. Same as ADI.

- **rhabdomyolysis**
acute, fulminating, potentially lethal disease of skeletal muscle that causes disintegration of striated muscle fibres as evidenced by myoglobin in the blood and urine.

- **rhinitis**
inflammation of the nasal mucosa.

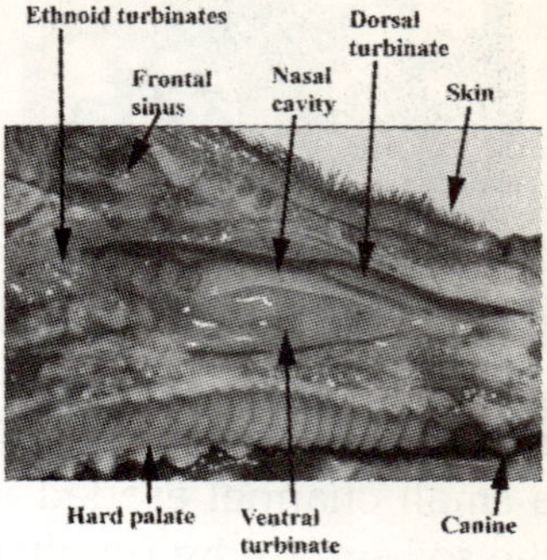

- **rhonchus**
harsh crepitation in the throat, often resembling snoring. See **crepitations**.

- **ribonucleic acid (RNA)**
RNA is the generic term for polynucleotides, similar to DNA but containing ribose in place of deoxyribose and uracil in place of thymine. These molecules are involved in the transfer of information from DNA, programming protein synthesis and maintaining ribosome structure. The 4 main types of RNA are heterogeneous nuclear RNA (hRNA), messenger RNA (mRNA), transfer RNA (tRNA) and ribosomal RNA (rRNA).

- **rill**
  a small channel eroded into the soil surface by runoff, rills can easily be smoothed out (obliterated) by normal tillage.

- **riparian rights**
  a doctrine of state water law under which a land owner is entitled to use the water on or bordering his property, including the right to prevent diversion or misuse of upstream waters. Riparian land is land that borders on surface water.

- **risk**
  possibility that a harmful event (death, injury or loss) arising from exposure to a chemical or physical agent may occur under specific conditions. See **excess lifetime risk, extra risk hazard.**

- **risk acceptance**
  decision that the risk associated with a given chemical exposure or an event leading to such exposure is low enough to be tolerated in order to gain associated benefits. See **acceptable risk**.

- **risk assessment**
  identification and quantification of the risk resulting from a specific use or occurrence of a chemical or physical agent, taking into account possible harmful effects on individual people or society of using the chemical or physical agent in the amount and manner proposed and all the possible routes of exposure. Quantification ideally requires the establishment of dose-effect and dose-response relationships in likely target individuals and

populations. See **exposure assessment, hazard identification, risk characterization, risk estimation, risk evaluation, risk identification, risk management, risk perçeption.**

- **risk assessment management process**
global term for the whole process from hazard identification to risk management. See **hazard identification, risk management**.

- **risk associated with a life time exposure**
probability of the occurrence of a specified undesirable event following exposure of an individual person from a given population to a specified substance at a defined level for the expected lifetime of the average member of that population.

- **risk aversion**
term used to describe the tendency of an individual person to avoid risk.

- **risk characterisation**
outcome of hazard identification and risk estimation applied to a specific use of a substance or occurrence of an environmental health hazard. The assessment requires quantitative data on the exposure of organisms or people at risk in the specific situation. The end product is a quantitative statement about the proportion of organisms or people affected in a target population. See **hazard identification, risk estimation.**

- **risk communication**
interpretation and communication of risk assessments in terms that are comprehensible to the general public or to others without specialist knowledge.

- **risk de minimis**
risk that is negligible and too small to be of societal concern (usually assumed to be a probability below 10^{-5} or 10^{-6}), can also mean 'virtually safe'. See **negligible risk.**

- **risk estimate**
a description of the probability that organisms exposed to a specified dose of chemical will develop an adverse response (e.g., cancer).

• **risk estimation**
assessment, with or without mathematical modelling, of the probability and nature of effects of exposure to a substance based on quantification of dose-effect and dose-response relationships for that substance and the population(s) and environmental components likely to be exposed and on assessment of the levels of potential exposure of people, organisms and environment at risk. See **risk evaluation, exposure assessment, hazard identification**.

• **risk evaluation**
establishment of a qualitative or quantitative relationship between risks and benefits, involving the complex process of determining the significance of the identified hazards and estimated risks to those organisms or people concerned with or affected by them. See **exposure evaluation, hazard identification, risk assessment, risk characterisation, risk estimation, risk identification, risk perception**.

• **risk factor**
a characteristic (e.g., race, sex, age, obesity) or variable (e.g., smoking, exposure) associated with increased chance of toxic effects. Some standard risk factors used in general risk assessment calculations include average breathing rates, average weight, and average human life span.

• **risk identification**
recognition of a potential hazard and definition of the factors required to assess the probability of exposure of organisms or people to that hazard and of harm resulting from such exposure.

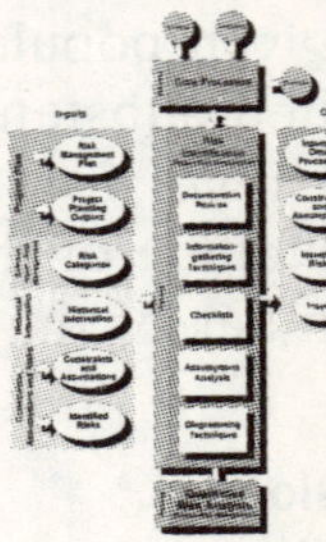

• **risk indicator**
see **risk marker**.

• **risk management**
decision-making process involving considerations of political, social, economic, and engineering

factors with relevant risk assessments relating to a potential hazard so as to develop, analyse, and compare regulatory options and to select the optimal regulatory response for safety from that hazard. Essentially risk management is the combination of three steps: risk evaluation, emission and exposure control and risk monitoring. See **emission and exposure control, risk evaluation, risk monitoring**.

- **risk marker**
  attribute that is associated with an increased probability of occurrence of a disease or other specified outcome and that can be used as an indicator of this increased risk: not necessarily a causal or pathogenic factor. See **risk indicator**.

- **risk monitoring**
  process of following up the decisions and actions within risk management in order to check whether the aims of reduced exposure and risk are achieved. See **monitoring, risk management**.

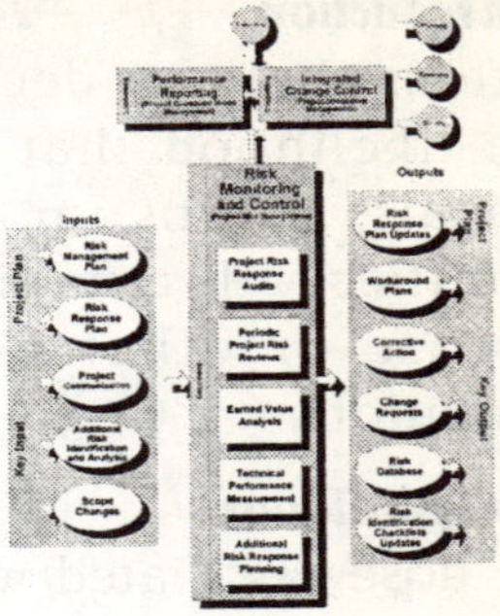

- **risk perception**
  subjective perception of the gravity or importance of the risk based on a person's knowledge of different risks and the moral, economic, and political judgement of their implications. See **risk evaluation.**

- **risk ratio**
  value obtained by dividing the probability of occurrence of a specific effect in one group by the probability of occurrence of the same effect in another group, or the value obtained by dividing the probability of occurrence of one potentially hazardous event by the probability of occurrence of another. Calculation of such ratios is used in choosing between options in risk management. See **risk management.**

- **risk reduction**
  actions that can decrease the likelihood that individuals, groups, or communities will experience disease or other health conditions.

- **risk specific dose**
  the dose associated with a specified risk level.

- **rodenticide**
  substance intended to kill rodents.

- **rotameter (rode-uh-me-ter)**
  a device used to measure the flow rate of gases and liquids. The gas or liquid being measured flows vertically up a tapered, calibrated tube. Inside the tube is a small ball or bullet-shaped float (it may rotate) that rises or falls depending on the flow rate. The flow rate may be read on a scale behind or on the tube by looking at the middle of the ball or at the widest part or top of the float.

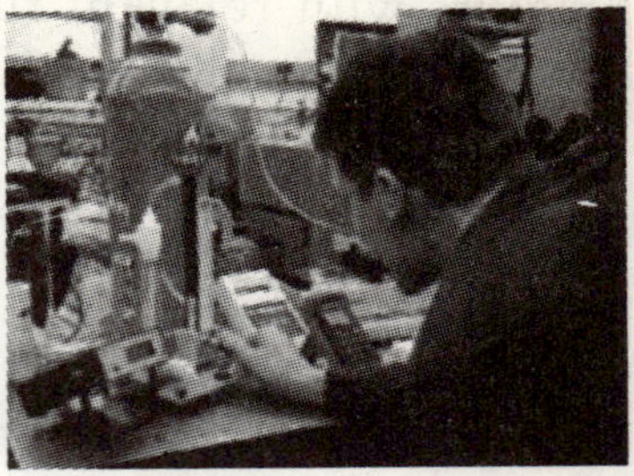

- **rotor**
  the rotating part of a machine. The rotor is surrounded by the stationary

(non-moving) parts (stator) of the machine.

- **route of exposure**
  means by which a toxic agent gains access to an organism by administration through the gastrointestinal tract (ingestion), lungs (inhalation), skin (topical), or by other routes such as intravenous, subcutaneous, intramuscular or intraperitoneal routes.

• **rubella**
a contagious viral infection characterised by a red rash, fever, and enlarged lymph nodes. Maternal infection in the early stage of pregnancy may result in congenital defects of the foetus. Blood tests are performed to detect immunity (antibodies) to rubella. Rubella is commonly known as German measles.

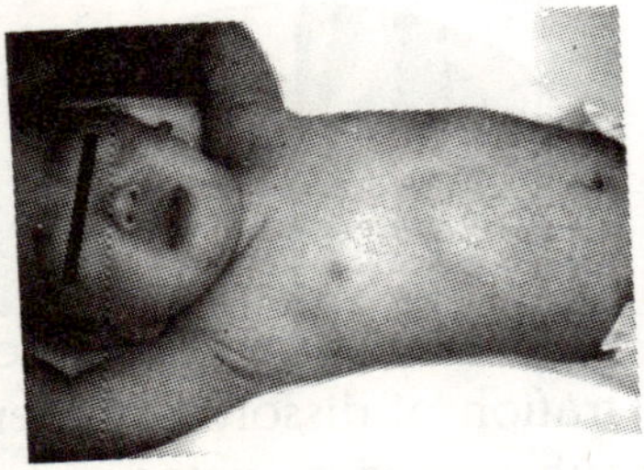

• **run-off**
water from rain, snow melt, or irrigation that flows over the ground surface and returns to streams. It can collect pollutants from air or land and carry them to the receiving waters.

• **sacrificial anode**
an easily corroded material deliberately installed in a pipe or tank. The intent of such an installation is to give up (sacrifice) this anode to corrosion while the water supply facilities remain relatively corrosion free.

• **safe**
strictly, free from harm or risk. Exposure to a chemical usually has some risk associated with it, although the risk may be very small. However, many people use the word safe to mean something that has a very low risk or one that is acceptable to them.

• **safe water**
water that does not contain harmful bacteria, or toxic materials or chemicals. Water may have taste and odour problems, colour and certain mineral problems and still be considered safe for drinking.

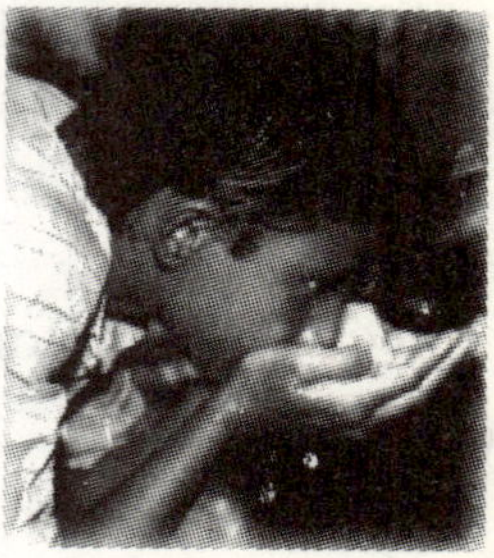

• **safe yield**
the annual quantity of water that can be taken from a source of supply over a period of years without deplet-

ing the source beyond its ability to be replenished naturally in 'wet years'.

- **safety**

1. reciprocal of risk: practical certainty that injury will not result from a hazard under defined conditions.
2. safety of a drug or other substance in the context of human health: the extent to which a substance may be used in the amount necessary for the intended purpose with a minimum risk of adverse health effects.
3. safety (toxicological): the high probability that injury will not result from exposure to a substance under defined conditions of quantity and manner of use, ideally controlled to minimise exposure. RT practical certainty, risk.

- **safety factor**

see **uncertainty factor**.

- **salicylates**

the drugs or compounds derived from salicylic acid — such as aspirin — that have fever reducing, anti-inflammatory, and pain relieving properties. Toxic concentrations may lead to nausea, impaired haemostasis, and electrolyte disturbance. The serum concentration of salicylate may be measured by a blood test.

- **salinity**

1. the relative concentration of dissolved salts, usually sodium chloride, in a given water.

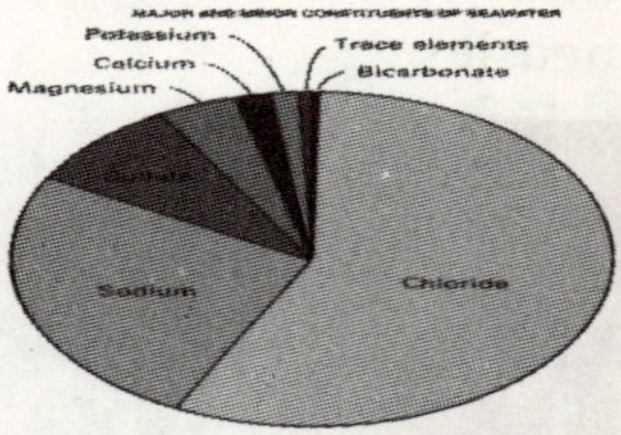

2. a measure of the concentration of dissolved mineral substances in water.

- **saluretic**

see **natriuretic**.

- **sample**

1. in statistics, a group of individuals often taken at random from a population for research purposes
2. one or more items taken from a population or a process and intended to provide information on the population or process.
3. portion of material selected from a larger quantity in some manner chosen so that

the portion is representative of the whole. See **biased sample**, **random sample**, **stratified sample**, **systematic sample**.

- **sample size**
  the number of units chosen from a population or an environment.

- **sampling**
  procedure used to obtain or constitute a sample. See **sample**

- **sampling error**
  part of the total estimation error of a parameter (or value of a property, such as concentration) caused by the random nature of the sample. See **sample**, **sampling**.

- **sand filters**
  devices that remove some suspended solids from sewage. Air and bacteria decompose additional wastes filtering through the sand so that cleaner water drains from the bed.

- **sand filtration**
  a process of filtering wastewater through sand. The wastewater trickles over the bed of sand where air and bacteria decompose the wastes. The clean water flows out through drains in the bottom of the bed. The sludge accumulating at the surface must periodically be removed from the bed.

- **sand**
  soil particles between 0.05 and 2 .0 mm in diameter.

- **sanitary landfill, landfilling**
  protecting the environment when disposing of solid waste. Waste is spread in thin layers, compacted by heavy machinery and covered with soil daily.

- **sanitary sewer**
  a sewer that transports only wastewaters (from domestic residences and/or indus-

tries) to a wastewater treatment plant.

- **sanitary survey**
  an on-site review of the water source, facilities, equipment, operation, and maintenance of a public water system for the purpose of evaluating the adequacy of the facilities for producing and distributing safe drinking water.

- **sanitary water**
  water discharged from restrooms, showers, food preparation facilities, or other non-industrial operations. Also known as 'grey water'.

- **saprophytes**
  organisms living on dead or decaying organic matter. They help natural decomposition of organic matter in water.

- **sarcoma**
  malignant tumour arising in a connective tissue and composed primarily of anaplastic cells resembling supportive tissue.

- **saturated zone**
  the area below the water table where all open spaces are filled with water.

- **saturator**
  a device which produces a fluoride solution for the fluoridation process. The device is usually a cylindrical container with granular sodium fluoride on the bottom. Water flows either upward or downward through the sodium fluoride to produce the fluoride solution.

- **saturnism**
  intoxication caused by lead. See **plumbism**.

- **scanning electron microscope**
  an electron microscope with a beam that scans (moves point-to-point) over the surface of a specimen, producing a lifelike, three dimensional view on film or screen. The magnification of the image is usually less than that of a non-scanning electron microscope. For visualisation by a scanning microscope, tissues do not require extensive preparation and handling.

- **schedule, pipe**
a sizing system of arbitrary numbers that specifies the I.D. (inside diameter) and O.D. (outside diameter) for each diameter pipe. This term is used for steel, wrought iron, and some types of plastic pipe. Also used to describe the strength of some types of plastic pipe.

- **sclerosis**
hardening of an organ or tissue, especially that due to excessive growth of fibrous tissue.

- **scotoma**
area of depressed vision within the visual field, surrounded by an area of less depressed or normal vision.

- **screening**
1. carrying out of a test or tests, examination(s) or procedure(s) in order to expose undetected abnormalities, unrecognised (incipient) diseases, or defects: examples are mass X-rays and cervical smears.
2. pharmacological or toxicological screening consists of a specified set of procedures to which a series of compounds is subjected to characterise pharmacological and toxicological properties and to establish dose-effect and dose-response relationships.

- **screening level**
decision limit or cut-off point at which a screening test is regarded as positive.

- **scrubbing**
a common method of reducing stack air emissions, removal of impurities by spraying a liquid that concentrates the impurities into waste.

- **secchi disc (seck-key)**
a flat, white disc lowered into the water by a rope until it is just barely visible. At this point, the depth of the disc from the water surface is the recorded Secchi disc transparency.

- **secondary metabolite**
product of biochemical processes other than the normal metabolic pathways, mostly produced in micro-organisms or plants after the phase of active growth and under conditions of nutrient deficiency.

• **secondary treatment**
the second step taken by a Sewage Treatment Works in which bacteria consume the organic parts of the waste. This treatment usually removes about 90% of all solids and oxygen-demanding substances.

• **secondary wavelength**
a wavelength chosen because the colour developed by the reaction is minimal. It is also usually the wavelength at which the common clinical interfering substances have little or no absorbance.

• **secondhand smoke**
see **sidestream smoke**.

• **secretion**
1. process by which a substance such as a hormone or enzyme produced in a cell is passed through a plasma membrane to the outside, for example the intestinal lumen or the blood (internal secretion)
2. solid, liquid or gaseous material passed from the inside of a cell through a plasma membrane to the outside as a result of cell activity.

• **sedative**
substance that exerts a soothing or tranquillising effect. See **anaesthetic, narcotic**.

• **sediment**
1. particulate material that usually lies below water.
2. formulated particulate material that is intended to lie below water in a test.
3. topsoil, sand, and minerals washed from the land into water, usually after rain or snow melt. Sediments collecting in rivers, reservoirs, and harbours can destroy fish and wildlife habitat and cloud the water so that sunlight cannot reach aquatic plants. Loss of topsoil from farming, mining, or building activities can be prevented through a variety of erosion-control techniques.

• **sediment yield**
the quantity of sediment arriving at a specific location.

• **sedimentation**
a water treatment process in which solid particles settle out of the water being

treated in a large clarifier or sedimentation basin

- **seepage**
the percolation of water through the soil from unlined channels, ditches, watercourses and water storage facilities.

- **seize up**
seize up occurs when an engine overheats and a part expands to the point where the engine will not run. Also called 'freezing'.

- **self-cleaning of water (in a reservoir)**
water purification by natural biological and physico-chemical processes.

- **self-monitoring of blood glucose**
a means of testing levels of blood glucose via a fingerstick blood sample and a blood glucose testing meter to determine the glucose result. Regular monitoring is an essential component of any diabetes management regimen.

- **self-purification of the atmosphere**
purification of the atmosphere from contaminants by natural biological and physico-chemical processes. See **contaminant**.

- **semen**
a thick, whitish fluid secreted by the prostate to carry sperm.

- **semichronic**
see **subchronic**.

- **semi-confined aquifer**
an aquifer that is partially confined by a soil layer (or layers) of low permeability through which recharge and discharge can occur.

- **sensibilisation**
see **sensitisation**.

- **sensitivity**
the extent to which a diagnostic test correctly detects an abnormality. In defining a laboratory test's accuracy, sensitivity indicates the frequency of positive test results in patients who have the specific disease being tested for.

- **sensitivity (in analytical chemistry)**
  extent to which a small change in concentration of an analyte can cause a large change in the related measurement.

- **sensitivity (of a screening test)**
  extent (usually expressed as a percentage) to which a method gives results that are free from false negatives. Fewer the false negatives, the greater the sensitivity. Quantitatively, sensitivity is the proportion of truly diseased persons in the screened population who are identified as diseased by the screening test.

- **sensitisation**
  immune process whereby individuals become hypersensitive to substances, pollen, dandruff, or other agents that make them develop a potentially harmful allergy when they are subsequently exposed to the sensitising material (allergen). See **allergy, hypersensitivity**.

- **sensor**
  an instrument that measures (senses) a physical condition or variable of interest. Floats and thermocouples are examples of sensors. Also called a primary element

- **sensory effect level**
  1. intensity, where the detection threshold level is defined as the lower limit of the perceived intensity range (by convention the lowest concentration that can be detected in 50 % of the cases in which it is present).
  2. quality, where the recognition threshold level is defined as the lowest concentration at which the sensory effect can be recognized correctly in 50 % of the cases.
  3. acceptability and annoyance, where the nuisance threshold level is defined as the concentration at which not more than a small proportion of the population, less than 5 %, experiences annoyance for a small part of the time, less than 2 %. Since annoyance will be influenced by a number of factors, a nuisance threshold level cannot be set on the basis of concentration

alone. See **nuisance threshold**.

- **septage**
  the liquid and semisolid contents removed by pumping from a septic tank.

- **septic (sep-tick)**
  a condition produced by bacteria when all oxygen supplies are depleted. If severe, bottom deposits and water turn black, give off foul odours, and the water has a greatly increased chlorine demand.

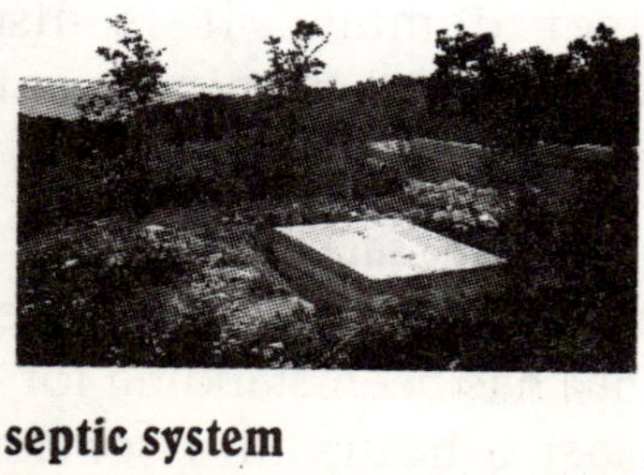

- **septic system**
  an onsite system designed to treat and dispose of domestic sewage. A typical septic system consists of a tank that receives wastes from a residence or business and a system of tile lines or a pit for disposal of the liquid effluent that remains after decomposition of the solids by bacteria in the tank.

- **septic tank**
  an underground tank to collect wastes from homes that are not connected to a municipal sewer system. Waste goes from the home to the tank and is decomposed by bacteria. Solids and dead bacteria settle to the bottom as sludge while the liquid portion flows into the ground through drains. While properly placed and maintained septic systems can effectively treat domestic wastewater, others are a major source of ground water and surface water pollution.

- **sequestration**
  a chemical completing (forming or joining together) of metallic cations (such as iron) with certain inorganic compounds, such as phosphate. Sequestration

prevents the precipitation of the metals (iron). Also see **chelation**.

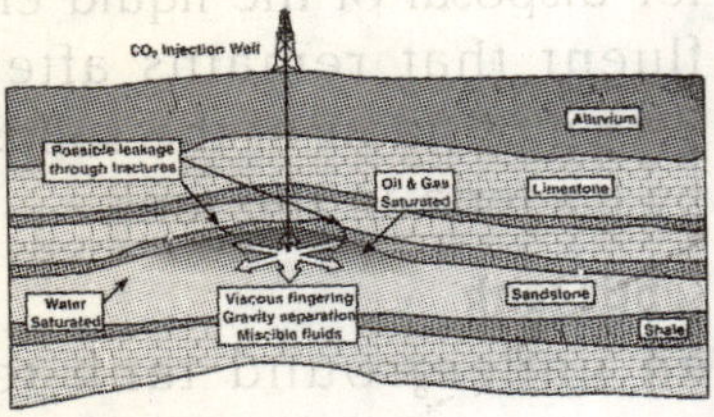

- **serologic testing**

an analysis of blood serum for evidence of infection by evaluating antigen-antibody reactions, used in acute and chronic stages of infection.

- **serum**

1. watery proteinaceous portion of the blood that remains after clotting. See **blood serum.**
2. clear watery fluid especially that moistening the surface of serous membranes or that exuded through inflammation of any of these membranes.
3. the clear, thin, watery portion of a liquid that can be separated from the more solid elements. The fluid portion of whole blood that separates after coagulation of the specimen. It is devoid of fibrinogen (a clotting protein). Serum may be separated out of whole blood by centrifugation.

- **serum glutamic oxaloacetic transaminase (SGOT)**

an enzyme present in all tissue, primarily in the liver, heart, and skeletal muscles. It is released into the bloodstream following cell death or injury. Elevated blood levels of SGOT may signal liver, heart, or skeletal muscle disease. It is also called aspartate aminotransferase (AST).

- **serum glutamic pyruvic transaminase (SGPT)**

an enzyme present in the same tissues as SGOT. Its appearance in serum is a marker of tissue damage similar to SGOT, but it is a more specific indicator of liver damage. It is also known as alanine aminotransferase (ALT).

- **service line sample**

a one-litre sample of water that has been standing for a least 6 hours in a service line.

- **service pipe**

the pipeline extending from the water main to the building served or to the consumers system.

- **set point**
  the position at which the control or controller is set. This is the same as the desired value of the process variable.
- **sewage**
  the used water and solids that flow from homes through sewers to a wastewater treatment plant. The preferred term is 'waste-water'.

- **sewage treatment works (STW)**
  a municipal or public service sewage treatment system.
- **sewer**
  a channel or conduit that carries wastewater and storm-water to a treatment plant or receiving waters. 'Sanitary' sewers carry household, industrial, and commercial waste. 'Storm' sewers carry runoff from rain or snow.

- **sewerage**
  the entire system of sewage collection, treatment, and disposal. Also applies to all effluent carried by sewers.
- **sewerage system**
  the network of sewers that carries sewage from point of origin to point of treatment.
- **sexually transmitted disease (STD)**
  a disease caused by a micro-organism that is passed to another person during sexual contact.
- **SGOT**
  see **serum glutamic oxalo-acetic transaminase**.
- **SGPT**
  see **serum glutamic pyruvic transaminase**.
- **shock load**
  the arrival at a water treating ment plant of raw water containing unusual amounts of algae, colloidal matter,

colour, suspended solids, turbidity, or other pollutants.

- **short term effect**
  see **acute effect**.
- **short term exposure limit (STEL)**
  average exposure that should not be exceeded at any time during a work day.
- **short-circuiting**
  a condition that occurs in tanks or basins when some of the water travels faster than the rest of the flowing water. This is usually undesirable since it may result in shorter contact, reaction, or settling times in comparison with the theoretical (calculated) or presumed detention times.
- **side-effect**
  action of a drug other than that desired for beneficial pharmacological effect.
- **siderosis**
  pneumoconiosis resulting from the inhalation of iron dust. See **pneumoconiosis**.

excess of iron in the urine, blood or tissues, characterised by haemosiderin.

granules in urine and iron deposits in tissues.

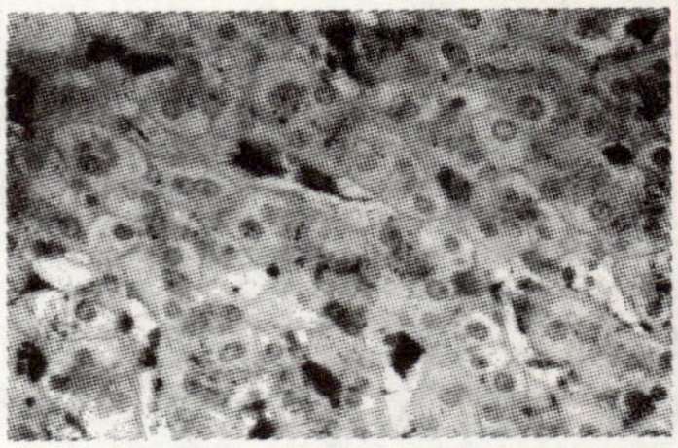

- **sidestream smoke**
  1. cloud of small particles and gases that is given off from the end of a burning tobacco product (cigarette, pipe, cigar) between puffs and is not directly inhaled by the smoker.
  2. the smoke that gives rise to passive inhalation on the part of bystanders.
- **sign**
  objective evidence of a disease, deformity or an effect induced by an agent, perceptible to an examining physician.
- **silicosis**
  pneumoconiosis resulting from inhalation of silica dust. See **pneumoconiosis**.

- **silt**
  soil particles between 0.05 and 0.002 millimetre in approximate diameter.

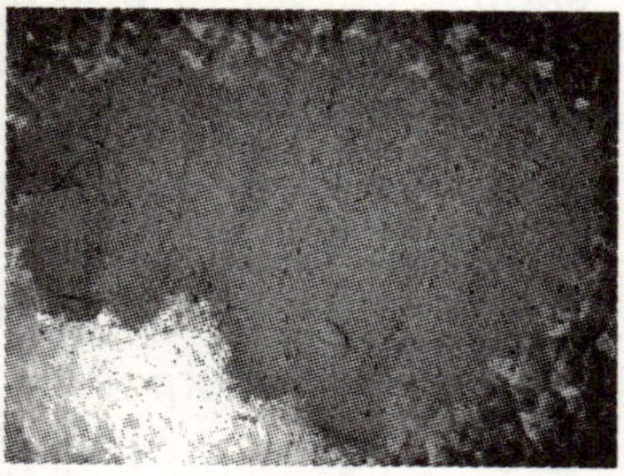

- **simulate**
  to reproduce the action of some process, usually on a smaller scale.
- **simulation test**
  procedure designed to predict the rate of biodegradation of a compound under relevant environmental conditions.
- **single family structure**
  a building constructed as a single-family residence that is currently used as either a residence or a place of business.
- **single-stage pump**
  a pump that has only one impeller. A multi-stage pump has more than one impeller.
- **sink**
  in environmental chemistry, an area or part of the environment in which, or a process by which, one or more pollutants is removed from the medium in which it is dispersed, for example - moist ground acts as a sink for sulfur dioxide in the air.
- **sister chromatid exchange**
  this is the process producing reciprocal exchange of DNA between the two DNA molecules of a replicating chromosome. See **deoxyribonucleic acid, chromosome**.
- **site inspection**
  a Department of Health visit to a site to evaluate the likelihood of human exposure to toxic chemicals, and to do an exposure assessment. See '**Exposure assessment**'

- **siting**
choosing a location for a facility.

- **skeletal fluorosis**
osteosclerosis due to fluoride.

- **slake**
to mix with water with a true chemical combination (hydrolysis) taking place, such as in the slaking of lime.

- **slimicide**
substance intended to kill slime-producing organisms (used on paper stock, water cooling systems, paving stones etc.).

- **slip correction factor**
a factor which allows slip flow behaviour to be calculated using continuum gas flow equations.

- **slip flow regime**
transition between free molecular flow and continuum gas flow.

- **slope factor**
value, in inverse concentration or dose units, derived from the slope of a dose-response curve. In practice, limited to carcinogenic effects with the curve assumed to be linear at low concentrations or doses. The product of the slope factor and the exposure is taken to reflect the probability of producing the related effect. See **concentration-effect curve, concentration-response curve, dose, dose-effect curve, dose-response curve**.

- **slope**
the slope or inclination of a trench bottom or a trench side wall is the ratio of the vertical distance to the horizontal distance or 'rise over run'.

- **slow sand filtration**
  means a process involving passage of raw water through a bed of sand at low velocity (generally less than 0.4 m/h) resulting in substantial particulate removal by physical and biological mechanisms.

- **sludge**
  the residue (solids and some water) produced as a result of raw water or wastewater treatment.

- **sludge blanket**
  an accumulation of solid matter hydrodynamically suspended with an enclosed body of water, such as in a clarifier.

- **slurry**
  a pumpable mixture of solids and fluid.

- **SMCLs**
  Secondary Maximum Contaminant Levels. Secondary MCLs for various water quality indicators are established to protect public welfare.

- **smog**
  dust, smoke, or chemical fumes that pollute the air and make hazy, unhealthy

  conditions (literally, the word is a blend of smoke and fog). Automobile, truck, bus, and other vehicle exhausts and particulates are usually trapped close to the ground, obscuring visibility and contributing to a number of respiratory problems.

- **smoke**
  an aerosol of solid or liquid particles usually resulting from incomplete combustion, usually in the form of aggregated very small primary particles. **See safety factor, modifying factor, no-observed-effect-level,**

no-observed-adverse-effect-level, reference dose.

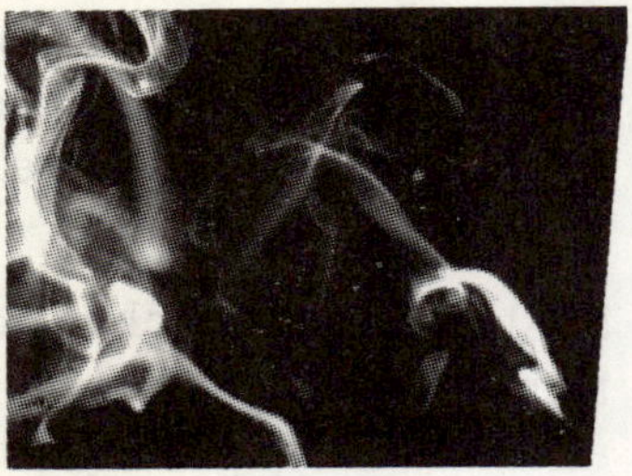

- **SNARL**
  Suggested No Adverse Response Level. The concentration of a chemical in water that is expected not to cause an adverse health effect.
- **Snell's law**
  fundamental principle in optics that the sines of the angles of incidence and refraction are in a constant ratio to one another.
- **societal risk**
  total probability of harm to a human population including also the probability of adverse health effects to descendants and the probability of disruption resulting from loss of services such as industrial plant or loss of material goods and electricity.

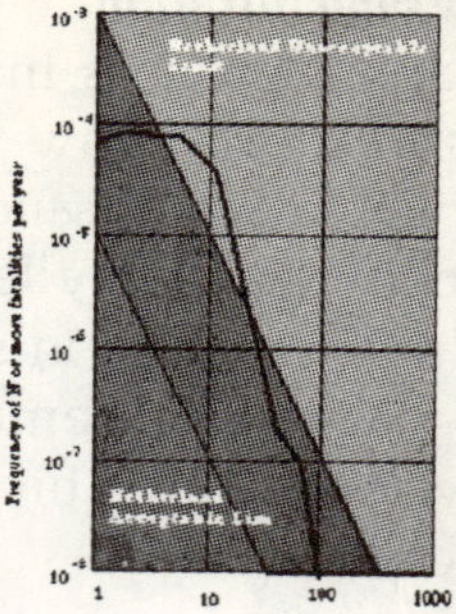

- **sodium (Na)**
  a chief electrolyte that regulates the amount of fluid inside and surrounding body cells. Measuring blood or urine levels of sodium will help to evaluate electrolyte and fluid balance disorders such as those that occur in edema, kidney, and liver or adrenal gland dysfunctions.
- **soft water**
  water having a low concentration of calcium and magnesium ions. Soft water is water having a hardness of 60 milligrams per litre or less.

- **software programs**
computer programs: the list of instructions that tell a computer how to perform a given task or tasks.

- **soil erodibility**
a measure of the soil's susceptibility to raindrop impact, runoff and other erosional processes.

- **soil profile**
a vertical section of the earth's highly weathered

upper surface often showing several distinct layers, or horizons.

- **soil structure**
the arrangement of soil particles into aggregates.

- **soil texture**
the proportions of soil particles (sand, silt, and clay) in a soil profile.

- **solder**
a metallic compound used to seal the joints between pipes. Until recently, most solder contained 50 percent lead. The use of lead solder containing more than 0.2% lead is now prohibited for pipes carrying potable water.

- **sole source aquifer**
an aquifer that supplies 50 percent or more of the drinking water of an area.

- **solenoid**
a magnetically (electrical coil) operated mechanical device. Solenoids can operate a small valve or a switch.

- **solid waste**
any solid, semi-solid, liquid, or contained gaseous materials discarded from industrial, commercial, mining, or agricultural operations, and from community activities. Solid waste includes garbage, construction debris, commercial refuge, sludge from water supply or waste

treatment plants, or air pollution control facilities, and other discarded materials.

- **solid waste management facility**
  any disposal or resource recovery system, any system, programme, or facility for resource conservation, any facility for the treatment of solid wastes.

- **solubility**
  the largest amount of a substance that can be dissolved in a given amount of a liquid, usually water. For a highly water-soluble compound, such as table salt, a lot can dissolve in water. Motor oil is only slightly soluble in water.

- **solution**
  a liquid mixture of dissolved substances. In a solution it is impossible to see all the separate parts.

- **solvent**
  a liquid capable of dissolving or dispersing another substance (for example, acetone or mineral spirits).

- **solvent abuse**
  deliberate inhalation (or drinking) of volatile solvents, in order to become intoxicated. See '**solvent sniffing**', '**glue sniffing**'.

- **somatic**
  1. pertaining to the body as opposed to the mind.
  2. pertaining to non-reproductive cells or tissues.
  3. pertaining to the framework of the body as opposed to the viscera.

- **sonography**
  a noninvasive diagnostic technique, also known as ultrasonography, in which high frequency sound waves are aimed at a selected body area or organ, with the resulting echoes being converted to images. Diagnostic ultrasound often is employed for obstetrical conditions.

- **soporific**
  substance producing sleep. See **anaesthetic**, **narcotic**, **sedative**.

• **sorption**
noncommittal term used instead of adsorption or absorption when it is difficult to discriminate experimentally between these two processes.

• **sounding tube**
a pipe or tube used for measuring the depths of water.

• **source of contamination**
the place where a hazardous substance comes from, such as a landfill, waste pond, incinerator, storage tank, or drum. A source of contamination is the first part of an exposure pathway.

• **source reduction**
the design, manufacture, purchase, or use of materials (such as products and packaging) to reduce the amount or toxicity of garbage generated. Source reduction can help reduce waste disposal and handling charges because the costs of recycling, municipal composting, land filling, and combustion are avoided,. Source reduction conserves resources and reduces pollution.

• **source separation**
organising materials by type (such as paper, metal, plastic, and glass) so that these items can be recycled instead of thrown away. For example, many of us separate these items from the rest of our household and office wastes. Industries also organise materials in this fashion.

• **space charge**
charged particles within an aerosol may interact with one another by virtue of the Coulomb-type forces which bring about mutual attraction or repulsion depending on whether particle charges are of the same or opposite polarities. Net result integrated over the whole cloud of particles is an internal electric field ultimately resulting in deposition onto solid surfaces.

• **span**
the scale or range of values an instrument is designed to measure. Also see **range**.

- **special populations**
  people who might be more sensitive or susceptible to exposure to hazardous substances because of factors such as age, occupation, sex, or behaviour (for example, cigarette smoking). Children, pregnant women, and older people are often considered special populations.
- **speciation**
  determination of the exact chemical form or compound in which an element occurs in a sample, for instance - determination of whether arsenic occurs in the form of trivalent or pentavalent ions or as part of an organic molecule, and the quantitative distribution of the different chemical forms that may coexist.
- **species**
  in biological systematics, group of organisms of common ancestry that are able to reproduce only among themselves and that are usually geographically distinct.
- **species differences in sensitivity**
  quantitative or qualitative differences of response to the action(s) of a potentially toxic substance on various species of living organisms. See **species-specific sensitivity**.
- **species-specific sensitivity**
  quantitative and qualitative features of response to the action(s) of a potentially toxic substance that are characteristic for particular species of living organism. See **species differences in sensitivity**.
- **specific conductance**
  a rapid method of estimating the dissolved-solids content of a water supply. The measurement indicates the capacity of a sample of water to carry an electrical current, which is related to the concentration of ionised substances in the water. Also called conductance.

- **specific death rate**
  death rate computed for a subpopulation of individual organisms or people having a specified characteristic or attribute, and named accordingly (for example, age-specific death rate, the number of deaths of persons of a specified age during a given period of time, divided by the total number of persons of that age in the population during that time).

- **specific gravity**
  weight of a particle, substance, or chemical solution in relation to the weight of water. Water has a specific gravity of 1.000 at 4 degrees C (39 degrees F). Particulates in raw water may have a specific gravity of 1.005 to 2.5.

- **specific organ or tissue toxins**
  toxin that affects only specific organs or tissues; these are called the target organs or target tissues.

- **specific pathogen free (SPF)**
  describing an animal removed from its mother under sterile conditions just prior to term and subsequently reared and kept under sterile conditions. See **germ-free animal.**

- **specific yield**
  the quantity of water that a unit volume of saturated permeable rock or soil will yield when drained by gravity. Specific yield may be expressed as a ratio or as a percentage by volume.

- **specificity**
  the ability of a diagnostic test to rule out the presence of a disease. In defining a laboratory test's accuracy, specificity indicates the frequency of negative test results in patients who do not have the specific disease being tested for.

- **specimen**
  specifically selected portion of any substance, material, organism (specifically tissue, blood, urine or faeces) or environmental medium assumed to be representative of the parent substance etc. at the time it is taken for the purpose of diagnosis, identification, study or demonstration.

- **spectrophotometric**
  pertaining to the estimation of colouring matter in a so-

lution using a device (spectrophotometer) that compares that colour to the full spectrum.

- **spina bifida**
  a congenital condition that is characterised by incomplete closure of the foetal vertebral arches. Defective closure of the bony canal that encases the spinal cord allows protrusion of portions of the brain and spinal cord through the defect.

- **spinal tap (spinal or lumbar puncture)**
  entrance by needle for removal of fluid from the spinal cavity, for diagnostic tests or to allow introduction of other fluids (anaesthetics, diagnostic aids, for example). Spinal or cerebrospinal fluid (CSF) bathes the brain, providing mechanical support and maintaining the chemical balance. Central nervous system diseases or injury can cause changes in CSF (elevated cell count, elevated protein, presence of tumour cells, bacteria, colour change, for example) that are of diagnostic importance.

- **spoil**
  excavated material such as soil from the trench of a water main.

- **spore**
  the reproductive body of an organism which is capable of giving rise to a new organism either directly or indirectly. A viable (able to live and grow) body regarded as the resting stage of an organism. A spore is usually more resistant to disinfectants and heat than most organisms.

- **spray**
  an aerosol of relatively large liquid droplets produced by the mechanical disruption of bulk liquid material, with sizes upwards of a few micrometres.

• **spreader**
agent used in some pesticide formulations to extend the even disposition of the active ingredient.

• **spring**
ground water seeping out of the earth where the water table intersects the ground surface.

• **spring line**
theoretical centre of a pipeline. Also, the guideline for laying a course of bricks.

• **stabilisation**
the process of changing an active substance into inert, harmless material, or physical activities at a site that act to limit the further spread of contamination without actual reduction of toxicity.

• **stability half-life (half-time)**
time required for the amount of a substance in a formulation to decrease, for any reason, by one-half (50 %).

• **stage**
a term used to describe the extent of cancer.

• **staging**
tests conducted to determine the stage of a cancer.

• **stakeholder**
a person, group, or community who has an interest in activities at a hazardous waste site.

• **standard**
1. that which is established as a measure or model to which others of a similar nature should conform.
2. technical specification, usually in the form of a document available to the public, drawn up with the consensus or general approval of all interests affected by it, based on the consolidated results of science, technology and experience, aimed at the promotion of optimum community benefits and approved by a body recognised on the national, regional or international level.

- **standard material (in analytical chemistry)**
  1. reference material (or calibration material) for which, for specified element concentrations, values are recommended by some official body.
  2. substance sufficiently well defined to be used for calibration and quality control of measurement techniques. See **reference material**.

- **standard sample**
  the aliquot of finished drinking water that is examined for the presence of coliform bacteria.

- **standard solution**
  a solution in which the exact concentration of a chemical or compound is known.

- **standard(ised) mortality (morbidity) ratio (SMR)**
  ratio of the number of events observed in the study group or population to the number of deaths expected if the study population had the same specific rates as the standard population, multiplied by 100.

- **standard**
  a physical or chemical quantity whose value is known exactly, and is used to calibrate or standardise instruments. Also **see reference**.

- **standardisation**
  1. making any substance, drug or other preparation conform to type or precisely defined characteristics.
  2. establishment of precisely defined characteristics, or precisely defined methods, for future reference.
  3. definition of precise procedures for administering, scoring and evaluating the results of a new method that is under development.

- **stannosis**
  pneumoconiosis resulting from inhalation of tin dust.

- **starters**
  devices used to start up large motors gradually to avoid severe mechanical shock to a driven machine and to prevent disturbance to the electrical lines (causing dimming and flickering of lights).

- **static head**
  when water is not moving, the vertical distance (in feet)

from a specific point to the water surface is the static head. (The static pressure in psi is the static head in feet times 0.433 psi/ft.) Also see **static pressure.**

- **static pressure**
when water is not moving, the vertical distance (in feet) from a specific point to the water surface is the static head. The static pressure in psi is the static head in feet times 0.433 psi/ft. Also see **static head**

- **static water depth**
the vertical distance in feet from the centreline of the pump discharge down to the surface level of the free pool while no water is being drawn from the pool or water table.

- **statistics**
a branch of mathematics that deals with collecting, reviewing, summarising, and interpreting data or information. Statistics are used to determine whether differences between study groups are meaningful.

- **stator**
that portion of a machine which contains the stationary (non-moving) parts that surround the moving parts (rotor).

- **sterilisation**
the removal or destruction of all micro-organisms, including pathogenic and other bacteria, vegetative forms and spores.

- **stethoscope**
an instrument used to magnify sounds and convey them to the ear.

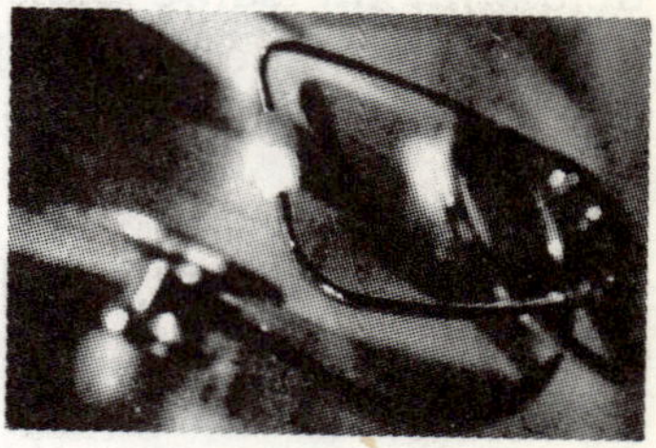

- **stochastic**
of, pertaining to or arising from chance and hence involving probability and obeying the laws of probability.

- **stochastic effect**
consequence for which the probability of occurrence depends on the absorbed dose. Hereditary effects and cancer induced by radiation are considered to be stochastic effects. The term 'stochastic' indicates that

the occurrence of effects so named, would be random. This means that, even for an individual, there is no threshold of dose below which the effect will not appear, and the chance of experiencing the effect increases with increasing dose. See **all-or-none effect, quantal effect**.

- **Stokes' diameter**
diameter of a spherical particle with the same density and settling velocity as the particle in question.

- **Stokes' law flow**
flow around a body under the influence of viscous, but not inertial forces.

- **Stokes' number**
ratio of a particle's stopping distance to a characteristic dimension. Generally used as an indicator of similitude in particle behaviour in a given aerosol flow configuration.

- **Stokes' regime**
condition for which Stokes' law applies.

- **stopping distance**
product of relaxation time and the initial particle velocity; an indicator of a particle's ability to adjust to directional changes in aerosol flow.

- **storm sewer**
a sewer that collects and transports surface runoff to a discharge point (infiltration basin, receiving stream, treatment plant).

- **stratification (in epidemiology)**
1. process of or result of separating a sample into several subsamples according to specified criteria such as age groups, socio-economic status, etc.
2. the formation of separate layers (of temperature, plant, or animal life) in a lake or reservoir. Each layer has similar characteristics such as all water in the layer has the same temperature. Also see **thermal stratification**.

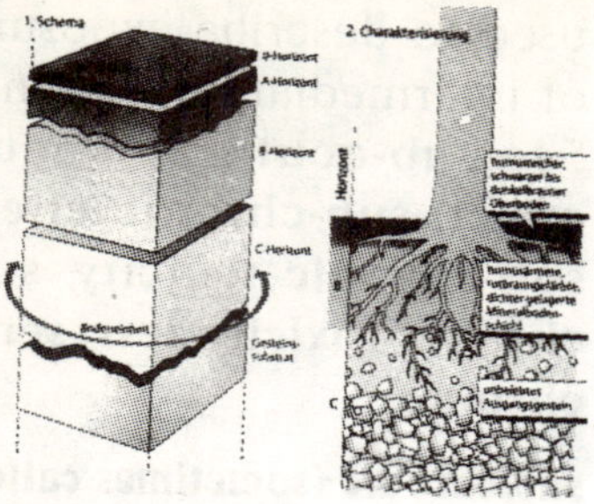

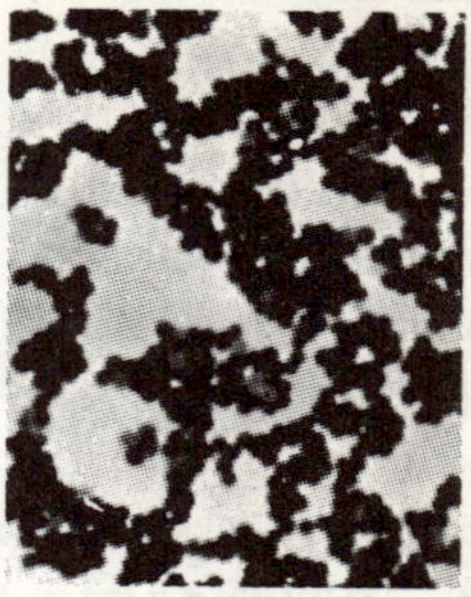

- **stratified sample**
  subset of a population selected according to some important characteristic. See **stratification.**

- **strep throat**
  infection of the throat and tonsils caused by group A streptococci. Symptoms include sore throat, fever, chills, and enlarged tonsils with white spots (pus). Strep throat or streptococcal sore throat may be detected by immunofluorescence staining of material from throat swabs. If proper treatment, including antibiotic administration, is not given, rheumatic fever can result.

- **streptococcus**
  a member of a family of coccoidal (round) bacteria that can cause a variety of external and internal infections.

- **strip cropping**
  a crop production system that involves planting alternating strips of row crops and close-growing forage crops; the forage strips intercept and slow runoff from the less protected row crop strips.

- **structure activity relationship (SAR)**
  the analysis of the dependence of the biological effects of a chemical upon its molecular structure produces a structure activity relationship. Molecular structure and biological activity are correlated by observing the results of systematic structural modification on defined biological endpoints. See **quantitative structure-activity relation (QSAR).**

• **sub-acute**
term used to describe a form of repeated exposure or administration usually occurring over about 21 days, not long enough to be called 'long-term' or 'chronic'. See **sub-chronic, sub-acute effect, sub-chronic effect, sub-chronic toxicity, sub-chronic toxicity test.**

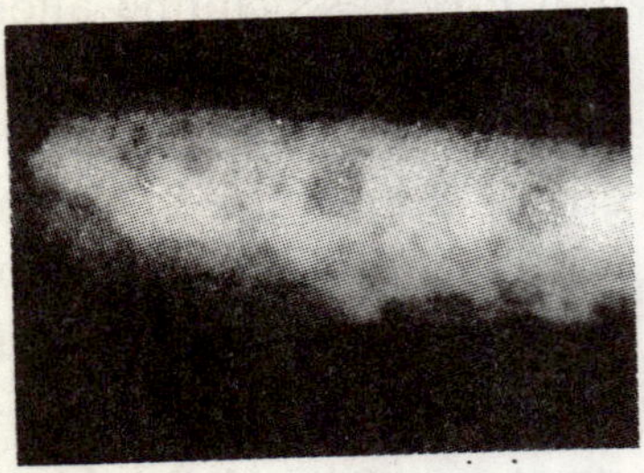

• **sub-acute (sub-chronic) toxicity**
adverse effects occurring as a result of repeated daily dosing of a chemical, or exposure to the chemical, for part of an organism's lifespan (usually not exceeding 10%). With experimental animals, the period of exposure may range from a few days to 6 months.

• **sub-chronic**
related to repeated dose exposure over a short period, usually about 10 % of the life span. An imprecise term used to describe exposures of intermediate duration. See **sub-acute, sub-acute effect, sub-chronic effect, sub-chronic toxicity, sub-chronic toxicity test, semi-chronic.**

• **sub-chronic (sometimes called sub-acute) effect**
biological change resulting from an environmental alteration lasting about 10 % of the lifetime of the test organism . In practice with experimental animals, such an effect is usually identified as resulting from multiple or continuous exposures occurring over 3 months (90 days). Sometimes a sub-chronic effect is distinguished from a sub-acute effect on the basis of its lasting for a much longer time. See sub-acute effect, sub-chronic toxicity, sub-chronic toxicity test.

• **sub-chronic toxicity test**
animal experiment serving to study the effects produced by the test material when administered in repeated doses (or continually in food, drinking-water, air)

over a period of up to about 90 days.

- **sub-chronic toxicity**
1. adverse effects resulting from repeated dosage or exposure to a substance over a short period, usually about 10 % of the life span.
2. the capacity to produce adverse effects following sub-chronic exposure.
See sub-acute, sub-chronic, sub-chronic effect, sub-chronic toxicity test.

- **sub-clinical effect**
biological change following exposure to an agent known to cause disease either before symptoms of the disease occur or when they are absent.

- **subjective environment**
surrounding conditions as perceived by persons living in these conditions.

- **submerged aquatic vegetation (SAV)**
aquatic vegetation, such as sea grasses, that cannot withstand excessive drying and therefore live with their leaves at or below the water surface. SAVs provide an important habitat for young fish and other aquatic organisms.

- **submergence**
the distance between the water surface and the media surface in a filter.

- **substance**
a chemical.

- **substance-specific applied research**
a programme of research designed to fill important data needs for specific hazardous substances. This research might include human studies or laboratory experiments to determine health effects resulting from exposure to a given hazardous substance.

- **sub-threshold dose**
see **non-effective dose.**

- **suction lift**
the negative pressure [in feet (meters) of water or inches (centimeters) of mercury vacuum] on the suc-

tion side of the pump. The pressure can be measured from the centreline of the pump down to (lift) the elevation of the hydraulic grade line on the suction side of the pump.

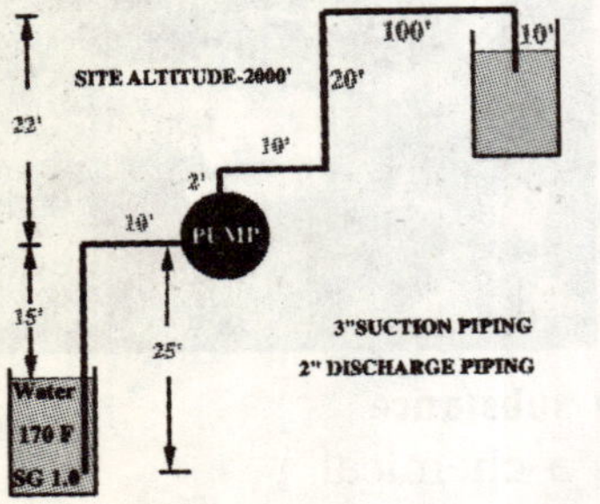

- **sudorific**
  substance that causes sweating.
- **sufficient evidence**
  sufficient evidence is a collection of facts and scientific references that is definite enough to establish that an adverse effect is caused by the agent in question. See carcinogenicity, limited evidence.
- **suggested no adverse response level (SNARL)**
  maximum dose or concentration that on current understanding is likely to be tolerated by an exposed organism without producing any harm.
- **summation (in neurophysiology)**
  process of addition of separate postsynaptic responses caused by stimuli that are adjacent in time and space. Excitation of a synapse evokes a graded potential change in the postsynaptic membrane that may be below the threshold required to trigger an impulse. If two or more such potentials are caused either nearly simultaneously, at different synapses on the same neurone (spatial summation), or in rapid succession at the same synapse (temporal summation), the summed response may be sufficient to trigger a postsynaptic impulse. Summation may occur between excitatory potentials, inhibitory potentials, or between an excitatory and an inhibitory potential.
- **superchlorination**
  chlorination with doses that are deliberately selected to produce free or combined residuals so large as to require de-chlorination.

• **supernatant**
liquid removed from settled sludge. Supernatant commonly refers to the liquid between the sludge on the bottom and the water surface of a basin or container.

• **supersaturated**
an unstable condition of a solution (water) in which the solution contains a substance at a concentration greater than the saturation concentration for the substance.

• **super-threshold dose**
see **toxic dose**.

• **supplier of water**
any person who owns or operates a public water system.

• **surface impoundment**
treatment, storage, or disposal of liquid hazardous wastes in ponds.

• **surface layer**
region of space comprising and adjoining the phase boundary between a solid and liquid phase, between a solid and gas phase, or between a liquid and gas phase within which properties of matter are significantly different from the values in the adjoining bulk phases. See **interfacial layer**.

• **surface loading**
one of the guidelines for the design of settling tanks and clarifiers in treatment plants. Used by operators to determine if tanks and clarifiers are hydraulically (flow) over- or underloaded. Also called overflow rate.

• **surface pump**
a mechanism for removing water or waste- water from a sump or wet well.

• **surface runoff**
precipitation, snow melt, or irrigation in excess of what can infiltrate the soil surface and be stored in small surface depressions. Runoff is a major transporter of non-point source pollutants.

• **surface water**
water on the surface of the earth, such as in lakes, rivers, streams, ponds, and springs

• **surge chamber**
a chamber or tank connected to a pipe and located at or near a valve that may quickly open or

close or a pump that may suddenly start or stop. When the flow of water in a pipe starts or stops quickly, the surge chamber allows water to flow into or out of the pipe and minimises any sudden positive or negative pressure waves or surges in the pipe.

- **surrogate**
relatively well studied toxicant whose properties are assumed to apply to an entire chemically and toxicologically related class, for example, benzo(a)pyrene data may be used as toxicologically equivalent to that for all carcinogenic polynuclear aromatic hydrocarbons. See **quantitative structure-activity relationship**.

- **surrogate species**
a species that is tested to estimate responses of other species, for which direct testing is impractical.

- **surveillance**
ongoing scrutiny, generally using methods distinguished by their practicability and uniformity, and frequently by their rapidity, rather than by complete accuracy. Its main purpose is to detect changes in trend or distribution in order to initiate investigative or control measures.

- **survey**
a systematic collection of information or data. A survey can be conducted to collect information from a

group of people or from the environment. Surveys of a group of people can be conducted by telephone, by mail, or in person. Some surveys are done by interviewing a group of people [see prevalence survey].

- **susceptibility**
condition of lacking the power to resist a particular disease or infection. Thus in susceptible people 'normal expected' results occur but with a lower exposure (or dose) than in the rest of the population.

- **suspended solids**
solids that either float on the surface or are suspended in water, wastewater, or other liquids.

- **sustainable agriculture**
environmentally friendly methods of farming that allow the production of crops or livestock without damage to the farm as an ecosystem, including effects on soil, water supplies, biodiversity, or other surrounding natural resources. The concept of sustainable agriculture is an 'intergenerational' one in which we pass on a conserved or improved natural resource base instead of one which has been depleted or polluted. Terms often associated with farms or ranches that are self-sustaining include 'low-input', 'organic', 'ecological', 'biodynamic', and 'permaculture'.

- **sympatholytic**
1. blocking transmission of impulses from the adrenergic (sympathetic) post-ganglionic fibres to effector organs or tissues.
2. agent that blocks transmission of impulses from the adrenergic (sympathetic) post-ganglionic fibres to effector organs or tissues. See **anti-adrenergic**.

- **sympathomimetic**
1. producing effects resembling those of impulses transmitted by the post-ganglionic fibres of the sympathetic nervous system.
2. agent that produces effects resembling those of impulses transmitted by the post-ganglionic fibres of the sympathetic nervous system. See **adrenergic**

- **symptom**
any subjective evidence of a disease or an effect induced by a substance as per-

ceived by the affected subject.

- **symptomatology**
general description of all of the signs and symptoms of exposure to a toxicant. Signs are the overt (observable) responses associated with exposure (such as convulsions, death, etc.) whereas symptoms are covert (subjective) responses (such as nausea, headache, etc.).

- **synapse**
functional junction between two neurones, where a nerve impulse is transmitted from one neurone to another.

- **synaptic transmission**
see **synapse**.

- **syndrome**
set of signs and symptoms occurring together and often characterising a particular disease-like state.

- **synergism**
pharmacological or toxicological interaction in which the combined biological effect of two or more substances is greater than expected on the basis of the simple summation of the toxicity of each of the individual substances.

- **synergistic effect**
biological effect following exposure simultaneously to two or more substances that is greater than the simple sum of the effects that occur following exposure to the substances separately. See **additive effect, antagonism, potentiation**.

- **system with a single service connection**
a system which supplies drinking water to consumers via a single service line.

- **systematic sample**
subset selected according to some simple rule such as specified date or alphabetic classification. See **biased sample, stratified sample**.

- **systemic**
relating to the body as a whole.

- **systemic effect**
consequence that is of either a generalised nature or that occurs at a site distant from the point of entry of a substance. A systemic effect

requires absorption and distribution of the substance in the body.

- **T cell**
  see **T lymphocyte**.
- **T lymphocyte**
  animal cell which possesses specific cell surface receptors through which it binds to foreign substances or organisms, or those which it identifies as foreign, and which initiates immune responses. See **B lymphocyte, immune response, lymphocyte**.
- **T.U.**
  toxic units. Expresses the relative toxicity of an effluent in such a manner that the larger the toxic unit value the more toxic the effluent. $T.U._{Ac} = 100/LC_{50}$. $T.U._{Chr} = 100/NOEC$. Examples $LC_{50} = 25\%$, $T.U._{Ac} = 4.0$ and $NOEC = 12.5\%$, $T.U._{Chr} = 8.0$. Permits with whole effluent toxicity limits (WET limits) often specify a T.U. value that may not be exceeded.
- **T3 thyrotoxicosis**
  a condition of hyperthyroidism where FT3, but not FT4, is elevated.
- **TAC (test acceptability criteria)**
  specific criteria for determining whether toxicity test results are acceptable . For example, a TAC for minimum control survival is typically specified in toxicity test methods.
- **tachy**
  prefix meaning rapid as in tachycardia and tachypnoea.
- **tachycardia**
  abnormally fast heartbeat. See **bradycardia**.
- **tachypnoea**
  abnormally fast breathing. See **bradypnoea**.
- **taeniacide**
  substance intended to kill tapeworms.
- **target (biological)**
  any organism, organ, tissue, cell or cell constituent that is subject to the action of a pollutant or other chemical, physical, or biological agent. See **receptor**.
- **target (of environmental pollution)**
  human being or any organism, organ tissue, cell, resource, or any constituent of the environment, living

or not, that is subject to the activity of a pollutant or other chemical or physical activity or other agent. See **receptor**.

- **target organ**
an organ (such as the liver or kidney) that is specifically affected by a toxic chemical.

- **target organ dose**
the target organ dose is the amount of a potentially toxic substance reaching the organ chiefly affected by that substance.

- **target population (epidemiology)**
1. collection of individuals, items, measurements, etc about which we want to make inferences: the term is sometimes used to indicate the population from which a sample is drawn and sometimes to denote any reference population about which inferences are required.
2. group of persons for whom an intervention is planned.

- **TCE**
see **trichloroethane**

- **TDS**
see **total dissolved solids**.

- **telemetry**
the electrical link between the transmitter and the receiver. Telephone lines are commonly used to serve as the electrical line.

- **technical directive**
see **standard**.

- **temperature sensor**
a device that opens and closes a switch in response to changes in the temperature This device might be a metal contact, or a thermocouple that generates minute electrical current proportional to the difference in heat, or a variable resistor whose value changes in response to changes in temperature. Also called a heat sensor

- **temporary acceptable daily intake**
value for the acceptable daily intake proposed for

guidance when data are sufficient to conclude that use of the substance is safe over the relatively short period of time required to generate and evaluate further safety data, but are insufficient to conclude that use of the substance is safe over a lifetime. A higher-than-normal safety factor is used when establishing a temporary ADI and an expiration date is established by which time appropriate data to resolve the safety issue should be available. See **acceptable daily intake**.

- **temporary maximum residue limit**
  temporary maximum residue limit is established for a specified, limited period when: Only a temporary acceptable daily intake has been established for the pesticide concerned. Although an acceptable daily intake has been established, the residue data are inadequate for firm maximum residue recommendations.

- **temporary safe reference action level (TSRAL)**
  this is the regulatory value defining the inhalation exposure level in the workplace which is safe for a short time but which should be reduced as soon as possible or appropriate respiratory protection employed.

- **ten-to-the-minus-sixth (10-6)**
  used in risk assessments to refer to the probability of risk. Literally means a chance of one in a million. Similarly, ten-to-the-minus-fifth means a probability of one in 100,000, and so on.

- **teratogen**
  this is the descriptor applied to any substance that can cause non-heritable birth defects.

- **teratogenesis**
  the induction of nonhereditary congenital malformations (birth defects) in a developing foetus by exogenous factors acting in the womb, interference with normal embryonic development.

- **teratogenic**
  anything which produces nonheritable birth defects is said to be teratogenic.

- **teratogenicity**
  potential to cause or the production of structural malformations or defects in offspring. See **developmental toxicity, embryotoxicity**.

- **terrace**
  a broad channel, bench, or embankment con- structed across the slope to intercept runoff and detain or channel it to protected outlets, thereby reducing erosion from agricultural areas.

- **tertiary treatment**
  an enhancement of normal sewage treatment operations to provide water of potable quality using further chemical and physical treatment.

- **testing of chemicals**
  1. in toxicology, evaluation of the therapeutic and potentially toxic effects of substances by their application through relevant routes of exposure with appropriate organisms or biological systems so as to relate effects to dose following application.
  2. in chemistry, qualitative or quantitative analysis by the application of one or more fixed methods and comparison of the results with established standards.

- **testosterone**
  a male sex hormone produced mostly by the testicles with a small amount produced by the adrenal glands. Testosterone stimulates a man's sexual activity and growth of other sex organs, including the prostate.

- **theophylline**
  a drug used as a relaxant of bronchial smooth muscle. Toxic levels can cause increased heartbeats, nausea, and seizures. A blood test is available to monitor levels of the drug.

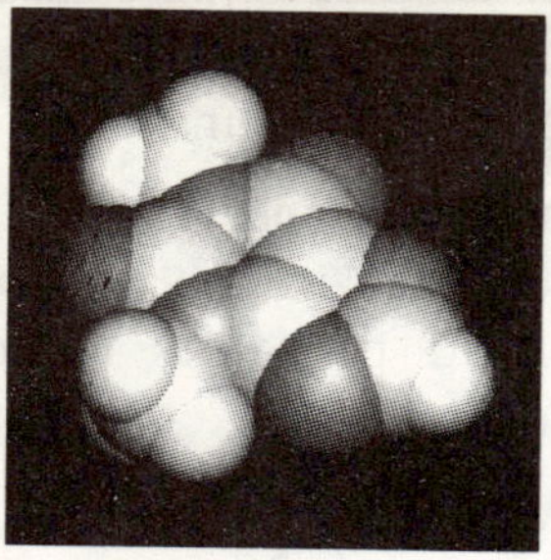

- **therapeutic index**
ratio between toxic and therapeutic doses (the higher the ratio, the greater the safety of the therapeutic dose).

- **thermal pollution**
the degradation of water quality by the introduction of a heated effluent. It is primarily the result of the discharge of cooling waters from industrial processes, particularly from electrical power generation. 'Waste heat' eventually results from virtually every energy conversion.

- **thermal stratification**
the formation of layers of different temperatures in a lake or reservoir. Also see **stratification**

- **thermocline**
the middle layer in a thermally stratified lake or reservoir. In this layer there is a rapid decrease in temperature with depth.

- **thermocouple**
a heat-sensing device made of two conductors of different metals joined at their ends. An electric current is produced when there is a difference in temperature between the ends.

- **THM**
see **trihalomethanes**

- **THM precursor**
see **precursor, THM.**

- **threshold**
  dose or exposure concentration below which an effect is not expected.

- **threshold limit value (TLV)**
  concentration in air of a substance to which it is believed that most workers can be exposed daily without adverse effect (the threshold between safe and dangerous concentrations). For most substances the value may be exceeded, to a certain extent, provided there are compensatory periods of exposure below the value during the workday (or in some cases the week). For a few substances (mainly those that produce a rapid response) the limit is given as a ceiling concentration that should never be exceeded.

- **threshold odour**
  the minimum odour of a water sample that can just be detected after successive dilutions with odourless water. Also called odour threshold.

- **threshold odour number**
  the greatest dilution of a sample with odour-free water that still yields a just-detectable odour.

- **thrombocytopenia**
  decrease in the number of blood platelets (thrombocytes).

- **thrust block**
  a mass of concrete or similar material appropriately placed around a pipe to prevent movement when the pipe is carrying water. Usually placed at bends and valve structures.

- **thyroglobulin (Tg)**
  a protein found in thyroid follicular cells that catalyses the production of T3 and T4 from MIT and DIT.

- **thyroglobulin antibodies (TgAB)**
  auto-antibodies produced by the body against its own thyroglobulin proteins. TgAb attacks the thyroid and disrupts thyroid function.

- **thyroid**
  a butterfly-shaped gland located at the base of the throat that controls metabolism through the secretion of T4 and T3 hormones.

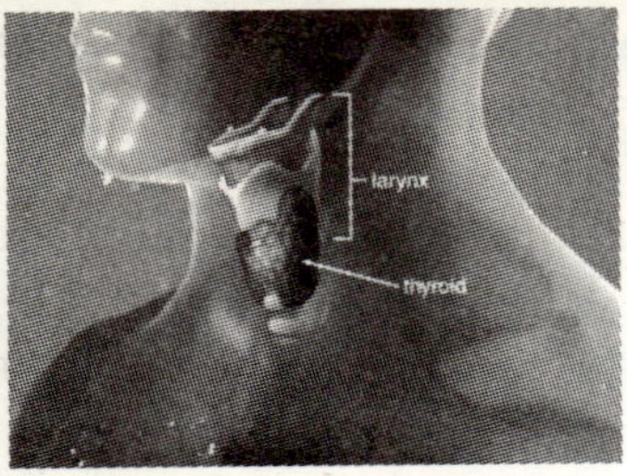

- **thyroid cancer**
  a rare thyroid disease characterised by uncontrolled tissue proliferation in the thyroid.
- **thyroid hormone binding ratio (THBR)**
  recommended nomenclature for T3 Uptake test. A ratio of the solid matrix uptake and the serum uptake.
- **thyroid nodules**
  masses (usually benign) that can form in the thyroid gland and may produce excessive thyroid hormone.
- **thyroid peroxidase (TPO)**
  a protein found in thyroid follicle cells that catalyses the iodination of T4 and T3 in thyroid hormone biosynthesis.
- **thyroid peroxidase antibodies (TPOAb)**
  auto-antibodies produced by the body against its own thyroid peroxidase enzymes. TPOAb attacks the thyroid and disrupts thyroid function.
- **thyroid stimulating hormone (TSH)**
  a hormone secreted from the anterior pituitary, that stimulates the thyroid to produce thyroid hormones T4 and T3. Elevated levels of thyroid hormones suppress TSH production by a classical feedback inhibition mechanism.
- **thyrotropin releasing hormone (TRH)**
  a tripeptide hormone produced by the hypothalamus which stimulates the anterior pituitary to produce TSH.
- **thyroxine (T4)**
  the primary hormone produced by the thyroid. Thyroxin circulates throughout the body primarily bound to carrier proteins. Free T4 is converted to triiodothyronine (T3) in peripheral tissues.
- **thyroxine binding globulin (TBG)**
  the major thyroid hormone transport protein in human serum.

- **thyroxine binding proteins (TBP)**
  three serum proteins with binding affinities for T4 and T3: thyroxine binding globulin (TBG), prealbumin, and albumin.

- **tidal volume**
  quantity of air or test gas that is inhaled and exhaled during one respiratory cycle.

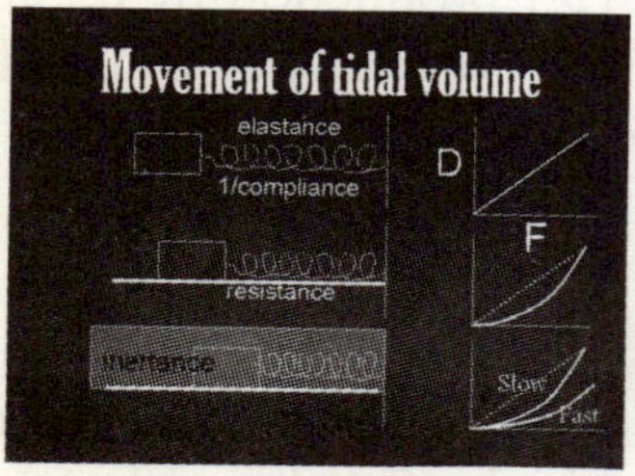

- **tillage**
  plowing, seedbed preparation, and cultivation practices.

- **time lag**
  the time required for processes and control systems to respond to a signal or to reach a desired level.

- **time weighted average concentration (TWA)**
  this is a regulatory value defining the concentration of a substance to which a person is exposed in ambient air, averaged over a period, usually 8 hours. For a person exposed to 0.1 mg m^{-3} for 6 hours and 0.2 mg m^{-3} for 2 hours, the 8 hour TWA is $(0.1 \times 6 + 0.2 \times 2) / 8$ which equals 0.125 mg m^{-3}

- **timer**
  a device for automatically starting or stopping a machine or other device at a given time.

- **time-weighted average**
  the average value of a parameter (e.g., concentration of a chemical in air) that varies over time.

- **time-weighted average exposure (TWAE) or concentration (TWAC)**
  concentration in the exposure medium at each measured time interval multiplied by that time interval and divided by the total time of observation for occupational exposure a working shift of eight hours is commonly used as the averaging time.

• **tinnitus**
continual noise in the ears, such as ringing, buzzing, roaring, or clicking.

• **tissue**
a group of cells organised to perform a specialised function.

• **tissue dose**
amount of a substance or physical agent (radiation) absorbed by a tissue.

• **tissue**
a group of similar cells.

• **titanic**
pertaining to tetanus, characterised by tonic muscle spasm.

• **titer**
the quantity of a substance or its maximal dilution that produces a measurable reaction with another specific substance. Antibody levels (titers) are measured to determine whether they are falling or rising.

• **titrate**
to titrate a sample, a chemical solution of known strength is added on a drop-by-drop basis until a certain colour change, precipitate, or pH change in the sample is observed (end point). Titration is the process of adding the chemical reagent in increments until completion of the reaction, as signalled by the end point

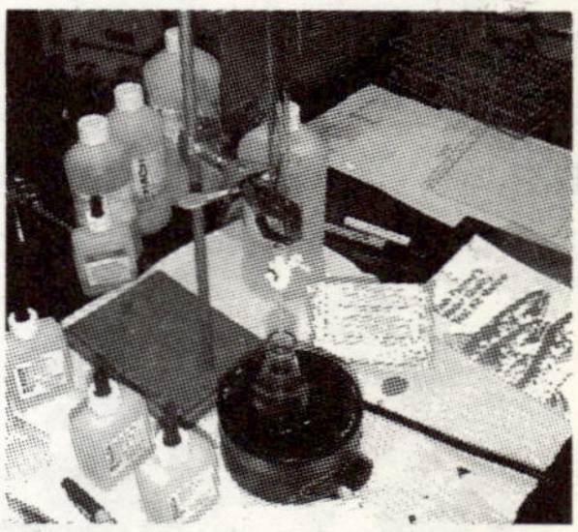

• **TNCWS**
see **transient non-community water system.**

• **tolerable daily intake (TDI)**
regulatory value equivalent to the acceptable daily intake. Unlike the ADI, the TDI is expressed in mg/person, assuming a body weight of 60 kg. TDI is normally used for food contaminants. See **acceptable daily intake.**

• **tolerable risk**
probability of suffering disease or injury that can, for the time being, be tolerated, taking into account the associated benefits, and assuming that the risk is minimised by appropriate control procedures. See **acceptable risk**.

- **tolerance**

1. adaptive state characterised by diminished effects of a particular dose of a substance. The process leading to tolerance is called 'adaptation'.
2. in food toxicology, dose that an individual can tolerate without showing an effect.

3. ability to experience exposure to potentially harmful amounts of a substance without showing an adverse effect.
4. ability of an organism to survive in the presence of a toxic substance: increased tolerance may be acquired by adaptation to constant exposure.
5. in immunology, state of specific immunological unresponsiveness.

- **tonic**

1. characterised by tension, especially muscular tension.
2. medical preparation that increases or restores normal muscular tension.

- **tonnage**

the amount of waste that a landfill accepts, usually expressed as tons per month. The rate at which a landfill accepts waste is limited by the landfill's permit.

- **too numerous to count**

the total number of bacterial colonies exceeds 200 on a 47-nm diameter membrane filter used for coliform detection.

- **topical**

pertaining to a particular area, as in a topical effect that involves only the area to which the causative substance has been applied.

- **topography**

the arrangement of hills and valleys in a geographic area.

- **total diet study**
  1. study designed to establish the pattern of pesticide residue intake by a person consuming a defined diet
  2. study undertaken to show the range and amount of various foodstuffs in the typical diet or to estimate the total amount of a specific substance in a typical diet.

- **total dissolved phosphorus**
  total phosphorus content of material that will pass through a filter of a specific size.

- **total dissolved solids (TDS)**
  the quantity of dissolved material in a given volume of water.

- **total dynamic head (TDH)**
  when a pump is lifting or pumping water, the vertical distance (in feet) from the elevation of the energy grade line on the suction side of the pump to the elevation of the energy grade line on the discharge side of the pump.

- **total maximum daily load (TMDL)**
  a determination of the amount of a pollutant, or property of a pollutant, from point, nonpoint, and natural background sources, including a margin of safety, that may be discharged into a water-quality limited waterbody .

- **total nitrogen**
  the sum of all nitrogen forms.

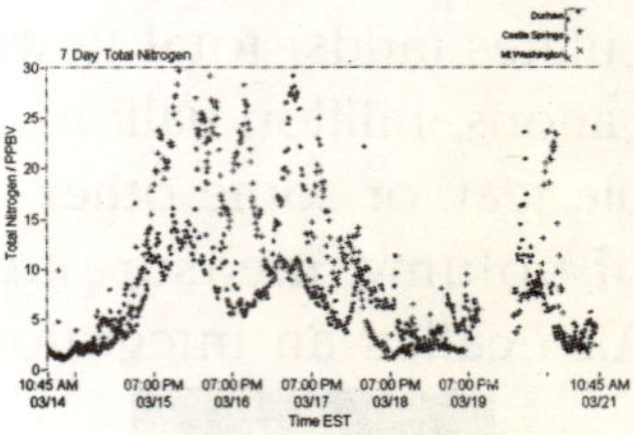

- **total particulate phosphorus**
  total phosphorus content of material retained on a filter of a specific size.

- **total phosphorus**
  the sum of all phosphorus forms.

- **total residual chlorine**
  the amount of available chlorine remaining after a given contact time. The sum of the combined available residual chlorine and the free available residual chlorine. Also see **residual chlorine**

- **total trihalomethanes (TTHM)**
  the sum of the concentration, in milligrams per litre, of the several trihalomethane compounds, rounded to two significant figures.

- **totalizer**
  a device or meter that continuously measures and calculates (adds) total flows in gallons, million gallons, cubic feet, or some other unit of volume measurement. Also called an integrator.

- **toxaphene (tox-uh-feen)**
  a chemical that causes adverse health effects in domestic water supplies and also is toxic to freshwater and marine aquatic life.

•

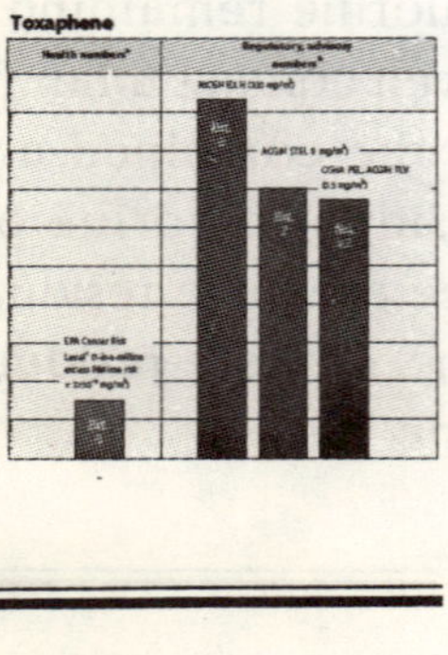

- **toxic**
  able to cause injury to living organisms as a result of physicochemical interaction.

- **toxic agent**
  chemical or physical (for example, radiation, heat, cold, microwaves) agents that, under certain circumstances of exposure, can cause harmful effects to living organisms.

- **toxic chemical**
  substances that can cause severe illness, poisoning, birth defects, disease, or death when ingested, inhaled, or absorbed by living organisms.

- **toxic cloud**
  an airborne mass of gases, vapours, fumes, or aerosols of toxic materials.

- **toxic dose**
  amount of a substance which produces intoxication without lethal outcome. See **super-threshold dose**.
- **toxic material**
  see **toxic substance**.
- **toxic nodular goitre**
  caused by autonomous production of thyroid hormone by solitary or multiple thyroid nodules.
- **toxic pollutants**
  materials contaminating the environment that cause death, disease. birth defects in organisms that ingest or absorb them. The quantities and length of exposure necessary to cause these effects can vary widely.

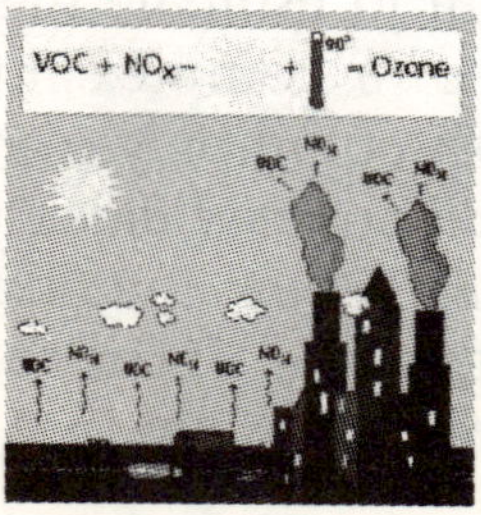

- **toxic substance**
  material causing injury to living organisms as a result of physicochemical interactions. See **chemical etiologic agent, poison, toxicant, toxic chemical, toxic material.**
- **toxification**
  metabolic conversion of a potentially toxic substance to a product that is more toxic.
- **toxicant**
  see **toxic substance**.

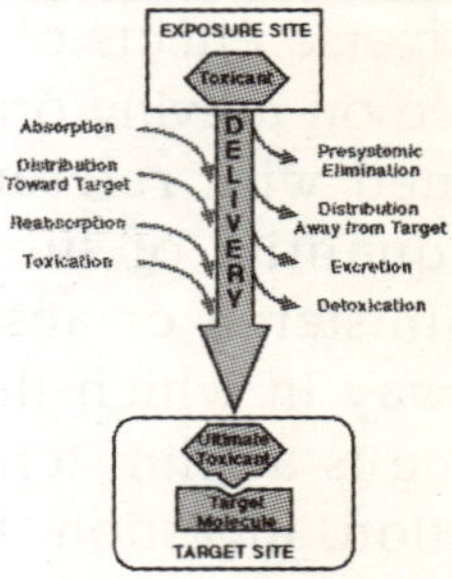

- **toxicity**
  1. capacity to cause injury to a living organism defined with reference to the quantity of substance administered or absorbed, the way in which the substance is administered (inhalation, ingestion, topical application, injection) and

distributed in time (single or repeated doses), the type and severity of injury, the time needed to produce the injury, the nature of the organism(s) affected and other relevant conditions.

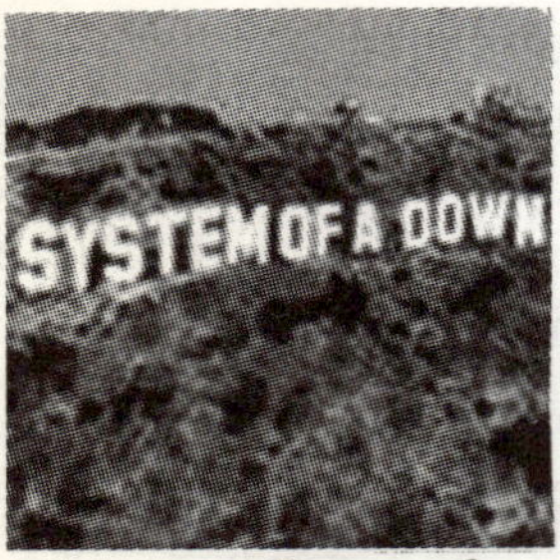

2. adverse effects of a substance on a living organism defined with reference to the quantity of substance administered or absorbed, the way in which the substance is administered (inhalation, ingestion, topical application, injection) and distributed in time (single or repeated doses), the type and severity of injury, the time needed to produce the injury, the nature of the organism(s) affected, and other relevant conditions.

3. measure of incompatibility of a substance with life: this quantity may be expressed as the reciprocal of the absolute value of median lethal dose ($1/LD_{50}$) or concentration ($1/LC_{50}$). See **acute toxicity, chronic toxicity, sub-acute toxicity, sub-chronic toxicity.**

4. the property of a chemical, or combination of chemicals, to adversely affect organisms, tissues or cells.

- **toxicity assessment**

this is the process of defining the nature of injuries that may be caused to an organism by exposure to a given chemical and the exposure concentration and time dependence of the chemically induced injuries. The aim of the assessment is to establish safe exposure concentration limits in relation to possible time of exposure.

- **toxicity equivalency factor (TEF)**

factor used in risk assessment to estimate the toxicity of a complex mixture, most commonly a mixture of chlorinated dibenzo-*p*-dioxins, furans and biphenyls. In this case, TEF is based on relative toxicity to 2,3,7,8 -

tetrachlorodibenzo-*p*-dioxin (TEF = 1).

- **toxicity equivalent (TEQ)**
contribution of a specified component (or components) to the toxicity of a mixture of related substances. The amount of substance (or substance concentration) of total toxicity equivalent is the sum of that for the components B, C.....N:
$S\ n(TEQ) = n(TEQ)_B + n(TEQ)_C + ...\ n(TEQ)_N$
Toxicity equivalent is most commonly used in relation to the reference toxicant 2,3,7,8-tetrachlorodibenzo-*p*-dioxin, 2,3,7,8-TCDD by means of the toxicity equivalency factor (TEF, *f*) which is 1 for the reference substance, hence:
$S\ n(TEQ) = f_B n_B + f_C n_C + ...\ f_N n_N$

- **toxicity identification evaluation (TIE)**
a set of procedures used to identify the specific compounds causing effluent toxicity.

- **toxicity reduction evaluation (TRE)**
a site-specific study conducted in a step-wise process designed to identify the causative agents of effluent toxicity, isolate the source of toxicity, evaluate the effectiveness of toxicity control options, and then confirm the reduction of effluent toxicity.

- **toxicity test**
experimental study of the adverse effects of exposure of a living organism to a substance for a defined duration under defined conditions. See **acute toxicity test**, **carcinogenicity test**, **chronic toxicity test**, **subchronic toxicity test**.

- **toxicity testing**
biological testing (usually with an invertebrate, fish, or small mammal) to determine the adverse effects, if any, of a chemical, compound, or effluent.

- **toxicodynamics**
process of interaction of potentially toxic substances with target sites, and the biochemical and physiological consequences leading to adverse effects. See adverse effect, pharmacodynamics, target.

- **toxicogenetics**
study of the influence of hereditary factors on the effects of potentially toxic substances on individual organisms. See **ecogenetics, pharmacogenetics, polymorphism**.

- **toxicokinetics**
process of the uptake of potentially toxic substances by the body, the bio-transformation they undergo, the distribution of the substances and their metabolites in the tissues, and the elimination of the substances and their metabolites from the body. Both the amounts and the concentrations of the substances and their metabolites are studied. The term has essentially the same meaning as pharmacokinetics, but the latter term should be restricted to the study of pharmaceutical substances. See **chemobiokinetics. See biotransformation, pharmacokinetics**.

- **toxicological data sheet**
document that gives in a uniform manner data relating to the toxicology of a substance, its production and application, properties and methods of identification. The data sheet may also include recommendations on protective measures. See **toxicological profile, toxicological dossier.**

- **toxicological profile**
a document that examines, summarises, and interprets information about a hazardous substance to determine harmful levels of exposure and associated health effects. A toxicological profile also identifies significant gaps in knowledge on the substance and describes areas where further research is needed.

- **toxicology**
scientific discipline involving the study of the actual or potential danger pre-

sented by the harmful effects of substances (poisons) on living organisms and ecosystems, of the relationship of such harmful effects to exposure, and of the mechanisms of action, diagnosis, prevention and treatment of intoxications. See **chemical toxicology**.

- **toxicometry**
term sometimes used to indicate a combination of investigative methods and techniques for making a quantitative assessment of toxicity and the hazards of potentially toxic substances.

- **toxicophobia**
morbid dread of poisons. See **chemophobia**.

- **toxicophoric (toxophoric) group**
structural moiety that upon metabolic activation exerts toxic effects. The presence of a toxicophoric group indicates only potential and not necessarily actual toxicity of a drug or other substances. See **toxogenic group**.

- **toxicovigilance**
active process of identification, investigation, and evaluation of various toxic effects in the community with a view to taking measures to reduce or control exposure(s) involving the substance(s) which produces these effects.

- **toxin**
poisonous substance produced by a biological organism such as a microbe, animal or plant. See **venom**.

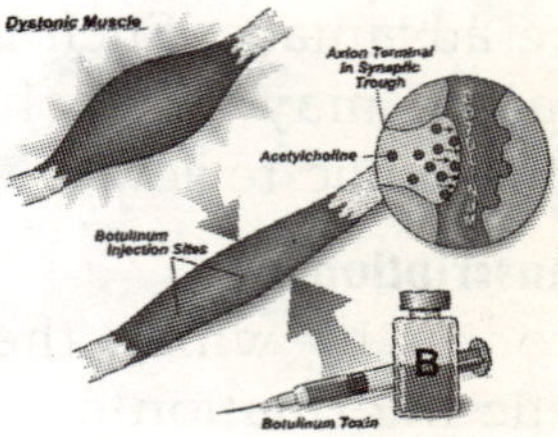

- **toxinology**
scientific discipline involving the study of the chemistry, biochemistry, pharmacology and toxicology of toxins. See **toxicology, toxin**.

- **toxogenic group**
see **toxicophoric group**.

- **tracer**
1. means by which something may be followed. For example a radioactive isotope may replace a stable chemical element in a toxic compound enabling the

toxicokinetics to be followed.
2. labelled member of a population used to measure certain properties of that population.

- **trade secret**
any confidential formula, pattern, process, device, information, or set of data that is used in a business to give the owner a competitive advantage. Such information may be excluded from public review.

- **transcription**
process by which the genetic information encoded in a linear sequence of nucleotides in one strand of DNA is copied into an exactly complementary sequence of RNA. See **reverse transcription**.

- **transducer (trans-due-sir)**
a device which senses some varying condition and converts it to an electrical signal for transmission to some other device (a receiver) for processing or decision making.

- **transformation**
1. alteration of a cell by incorporation of foreign genetic material and its subsequent expression in a new phenotype. See **phenotype.**
2. conversion of cells growing normally to a state of rapid division in culture resembling that of a tumour.
3. chemical modification of substances in the environment.

- **transformation (neoplastic)**
neoplastic transformation is the conversion of normal cells into tumour cells (see below). Frequently this is the result of a genetic change (mutagenesis) and the same term is used to describe the genetic modification of bacteria for use in biotechnology. See **epigenetic changes, genetic toxicology, genotoxic, mutation, mutagenesis, tumour.**.

- **transgenic**
adjective used to describe animals carrying a gene introduced by micro-injecting

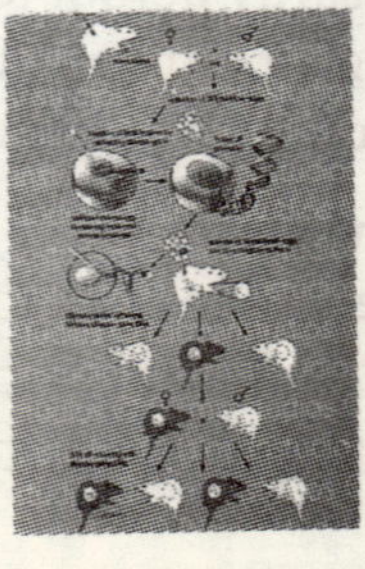

DNA into the nucleus of the fertilized egg.

- **transient water system**
  a non-community water system that does not serve 25 of the same non-resident persons per day for more than six months per year. Also called a transient non-community water system (TNCWS).

- **transmission lines**
  pipelines that transport raw water from its source to a water treatment plant. After treatment, water is usually pumped into pipelines (transmission lines) that are connected to a distribution grid system.

- **transmissivity**
  the ability of an aquifer to transmit water.

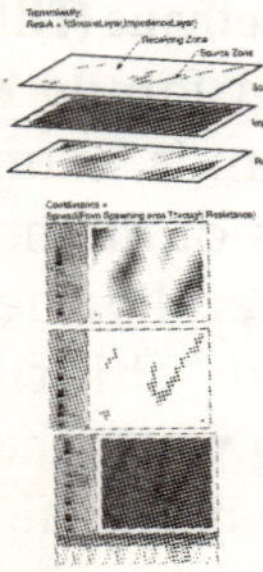

- **transpiration (tran-spur-ray-shun)**
  the process by which water vapour is released to the atmosphere by living plants.

- **treatability**
  in relation to waste water, the amenability of substances to removal without adversely affecting the normal operation of biological treatment processes (such as a sewage treatment plant).

- **treated wastewater**
  wastewater that has been subjected to one or more physical, chemical, and biological processes to reduce its pollution.

- **treatment, storage, and disposal facility (TSD)**
  refers to any facility which treats, stores, or disposes of hazardous wastes.

- **tremie (treh-me)**
  a device used to place concrete or grout under water.

- **triage**
  assessment of sick, wounded and injured persons follow-

ing a disaster to determine priority needs for efficient use of available medical facilities.

- **trickling filter**
  a biological treatment device. Wastewater is trickled over a bed of stones covered with bacterial growth, the bacteria break down the organic wastes in the sewage and produce cleaner water.

- **tricyclic antidepressants**
  widely used drugs for the treatment of depression. Drug overdose may cause severe toxic effects. The dexamethasone suppression test may aid in the selection of patients who might respond well to antidepressant treatment. Blood and urine tests may be used to determine levels of these drugs in the body.

- **triglyceride**
  a major food fat and blood lipid (fatty substance) produced by the liver from fatty acids. Excess triglycerides are stored in adipose tissue and are used to provide energy. Elevated blood levels of triglycerides and of cholesterol may be associated with an increased risk for atherosclerosis and related disorders, such as coronary artery disease, heart attack, and hypertension.

Triglyceride

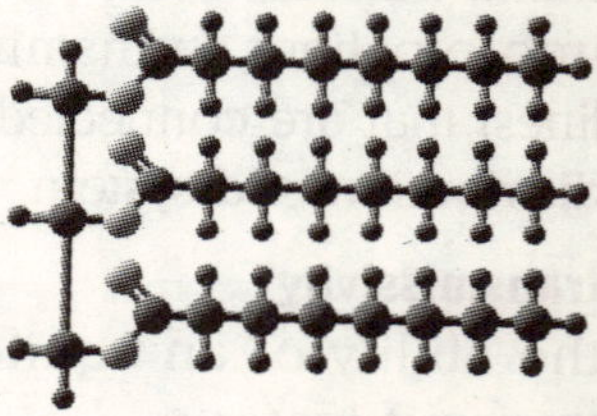

- **triholomethane (THM)**
  one of a family of organic compounds, named as derivatives of methane. THM's are generally the by-product from chlorination of drinking water that contains organic material.

- **triiodothyronine (T3)**
  a thyroid hormone primarily converted from T4, though some excretion occurs from the thyroid di-

rectly. T3 is the metabolically active hormone.

- **trohoc**

this term is regarded with disapproval by the majority of epidemiologists. It is used to describe an epidemiological study which starts with the outcome and looks backward for the causes.

- **trophic level**

amount of energy in terms of food that an organism needs. Organisms not needing organic food, such as plants, are said to be on a low trophic level, whereas predator species needing food of high energy content are said to be on a high trophic level. The trophic level indicates the level of the organism in the food chain.

- **tube settler**

a device that uses bundles of small bore (2 to 3 inches or 50 to 75 mm) tubes installed on an incline as an aid to sedimentation. The tubes may come in a variety of shapes including circular and rectangular. As water rises within the tubes, settling solids fall to the tube surface. As the sludge (from the settled solids) in the tube gains weight, it moves down the tubes and settles to the bottom of the basin for removal by conventional sludge collection means. Tube settlers are sometimes installed in sedimentation basins and clarifiers to improve particle removal.

- **tubercle**

a protective crust of corrosion products (rust) which builds up over a pit caused by the loss of metal due to corrosion.

- **tuberculation**

the development or formation of small mounds of corrosion products (rust) on the inside of iron pipe. These mounds (tubercules) increase the roughness of the inside of the pipe thus increasing resistance to water flow.

- **tumour**

an abnormal mass of tissue that results from excessive cell division that is uncontrolled and progressive. Tumours perform no useful body function. Tumours can be either benign (not cancer) or malignant (cancer).

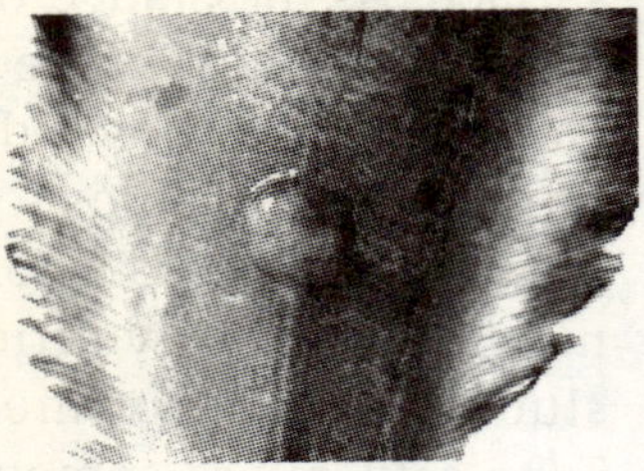

- **tumourigenic**
  able to cause tumours.

- **tumour progression**
  sequence of changes by which a benign tumour develops from the initial lesion to a malignant stage.

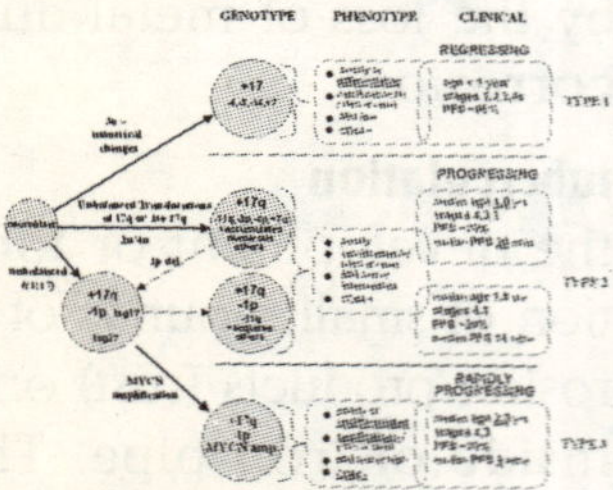

- **t-uptake**
  a measurement of the total binding capacity of the thyroxine binding proteins for thyroid hormones.

- **turbid**
  having a cloudy or muddy appearance.

- **turbidimeter**
  a device that measures the amount of suspended solids in a liquid.

- **turbidity**
  the cloudy appearance of water caused by the presence of suspended and colloidal matter. In the waterworks field, a turbidity measurement is used to indicate the clarity of water. Technically, turbidity is an optical property of the water based

  on the amount of light reflected by suspended particles. Turbidity cannot be directly equated to suspended solids because white particles reflect more light than dark-coloured particles and many small particles will reflect more light than an equivalent large particle.

- **Turner syndrome**
  a chromosomal abnormality in women in whom only one X chromosome is present instead of the normal two. The condition is characterised by growth abnormalities such as dwarfism and underdevel-

oped reproductive organs. Learning problems and mild mental retardation may be present. A smear of scrapings from inside the mouth may show evidence of this condition. Culture and chromosomal analysis of foetal cells from amniotic fluid can forecast this syndrome.

- **turnover time**
  see **mean life**.

- **TWS**
  see **transient water system**.

- **type 1 diabetes**
  a condition characterised by abrupt onset of symptoms, insulinopenia, dependence on exogenous insulin to sustain life and a tendency to develop ketoacidosis.

- **type 2 diabetes**
  a disorder usually characterised by a gradual onset with minimal or no symptoms of metabolic disturbance, exogenous insulin is not required to prevent ketonuria and ketoacidosis. Dietary control and exercise with or without oral diabetes medications is usually effective.

- **type i error (alpha)**
  the rejection of the null hypothesis (H_o) when it is, in fact, true (i.e., determining that the effluent is toxic when the effluent is not toxic) .

- **type ii error (beta)**
  the acceptance of the null hypothesis (H_o) when it is not true (i.e., determining that the effluent is not toxic when the effluent is toxic).

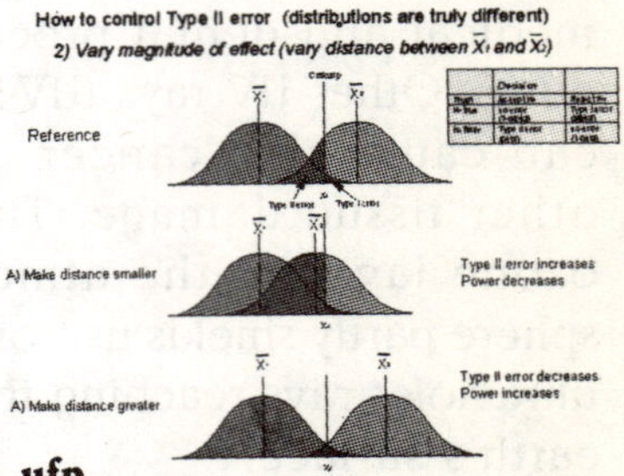

- **ufp**
  ultrafine particles, particles less than 100 nm in diameter that are a variable and heterogeneous component of environmental air pollution (PM10). They are derived from primary combustion sources.

- **ulcer**
  defect, often associated with inflammation, occurring locally or at the surface of an

organ or tissue owing to sloughing of necrotic tissue.

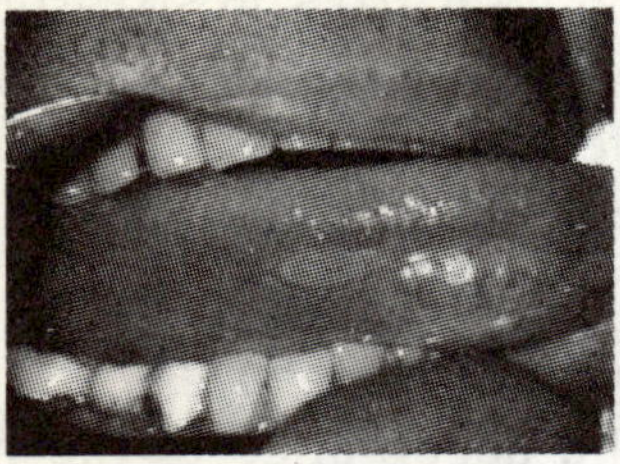

- **ultraviolet rays**

radiation from the sun in the invisible portion of the spectrum. Some UV rays (UV-A) enhance plant life and are useful in certain medical and dental procedures. Other UV rays (UV-B) can cause skin cancer or other tissue damage. The ozone layer in the atmosphere partly shields us from ultraviolet rays reaching the earth's surface.

- **uncertainty (safety) factor (UF)**

this term may be used in either of two ways depending upon the context

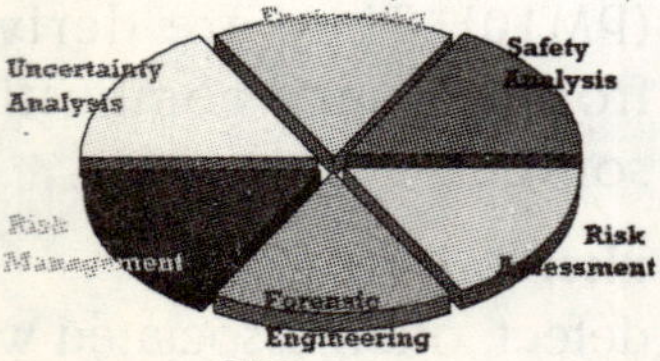

1. mathematical expression of uncertainty applied to data that are used to protect populations from hazards which cannot be assessed with high precision.

2. with regard to food additives and contaminants, a factor applied to the no-observed effect level to derive acceptable daily intake (ADI) (the no-observed-effect level is divided by the safety factor to calculate the ADI). The value of the safety factor depends on the nature of the toxic effect, the size and type of the population to be protected, and the quality of the toxicological information available.

- **unconfined aquifer**

an aquifer containing water that is not under pressure. The water level in a well is the same as the water table outside the well.

- **unidirectional**

refers to one-way communication between an instrument and a computer.

- **unit cancer risk**

estimate of the lifetime risk caused by each unit of exposure in the low exposure region.

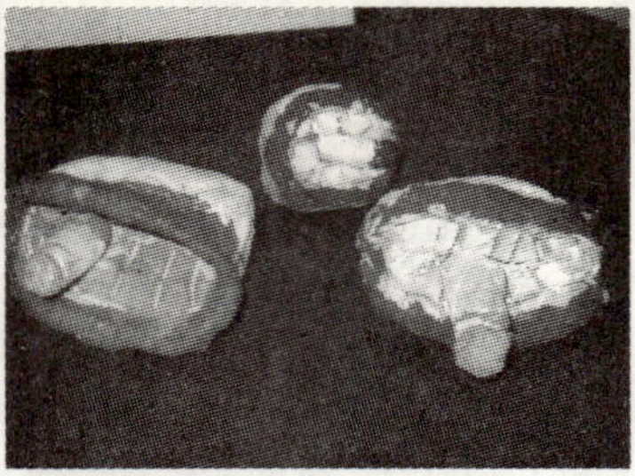

- **unit hydrograph**
the hydrograph of one inch of storm runoff generated by a rainstorm of fairly uniform intensity within a specific period of time.

- **unit risk**
incremental upper-boundary lifetime risk estimated to result from lifetime exposure to an agent if it is in air at a concentration of 1 mg/m^3 or in the water at a concentration of 1 mg/l.

- **unsaturated zone**
the area between the land surface and water table in which the pore spaces are only partially filled with water. Also called 'zone of aeration'.

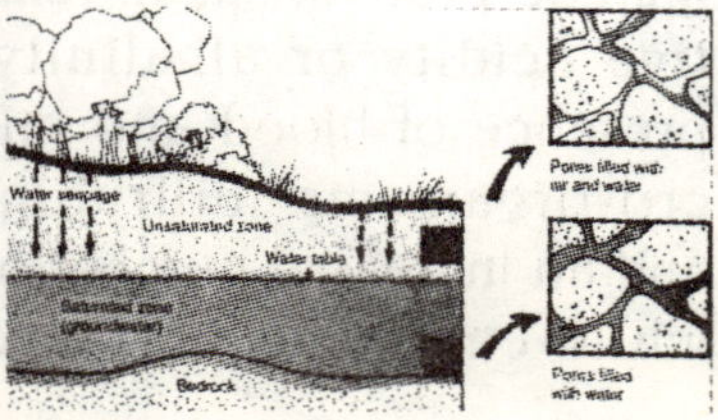

- **upper boundary**
estimate of the plausible upper limit to the true value of a quantity. This is usually not a statistical confidence limit.

- **upper-bound estimate**
estimate not likely to be lower than the true risk.

- **uptake**
entry of a substance into the body, into an organ, into a tissue, into a cell, or into the body fluids by passage through a membrane or by other means. See **absorption**.

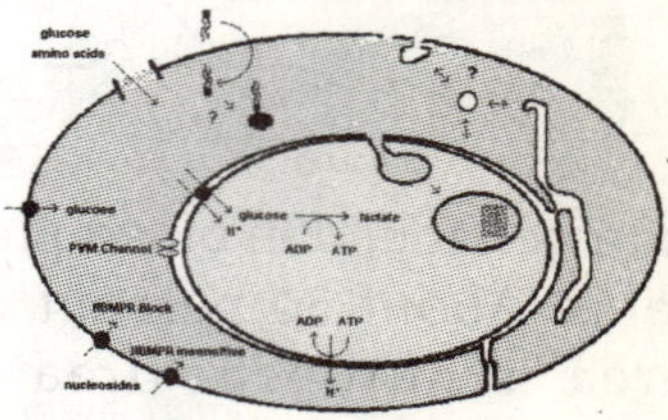

- **urate oxidase**
this is the hepatic peroxisomal enzyme which catalyses the oxygen mediated conversion of uric acid into allantoin. See **peroxisome**.

- **urban runoff**
stormwater from city streets and adjacent domestic or commercial properties that may carry pollutants of vari-

ous kinds into the sewer systems and/or receiving waters.

- **urea nitrogen**
the nitrogen that is found in urea (an end product of protein metabolism) which is recognised as different from

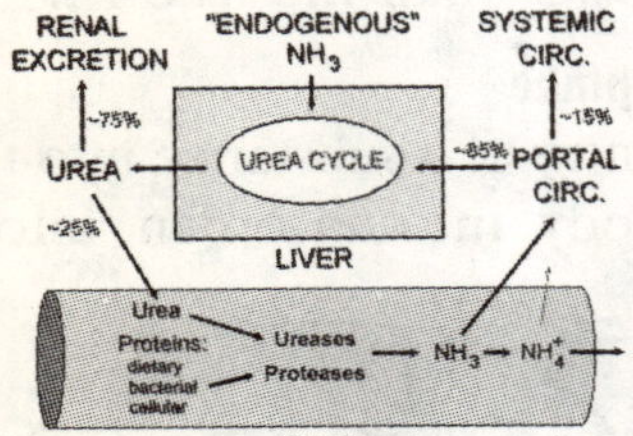

nitrogen found in blood proteins. About 50 percent of urea is nitrogen. Urea is normally present in blood and excreted in urine. Increased blood levels may be found in kidney disease, and decreased amounts are associated with liver disease.

- **urethra**
the canal that carries urine from the bladder and semen from the sex glands to the outside of the body.

- **urgent public health hazard**
sites where short-term exposures (less than 1 year) to hazardous substances or conditions could result in harmful health effects that require rapid intervention.

- **uric acid**
a product of protein metabolism. Measurement of uric acid in the blood is most commonly performed

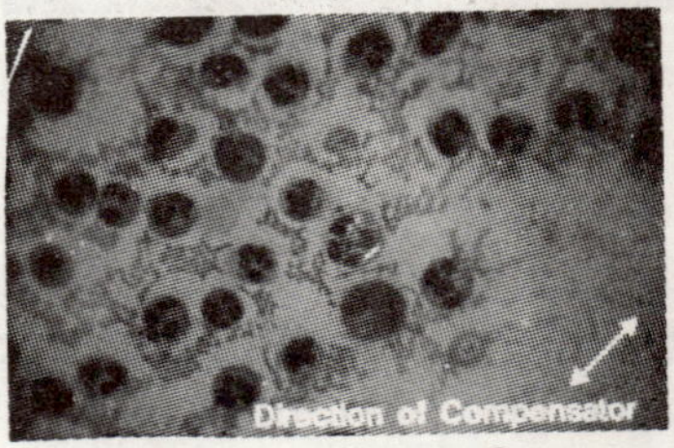

to evaluate gout. Gout is a metabolic disease marked by an accumulation of uric acid in the blood and of its salts in joints. This causes pain and swelling in the joints, especially in the great toe.

- **urinalysis**
the physical, chemical, and microscopic analysis of urine to detect changes in important characteristics and values that may indicate diseases or disorders. These include protein, sugar, relative acidity or alkalinity, presence of blood and micro-organisms, bilirubin, and an increased number of white cells. Special urine

tests, such as tests for pregnancy, are not performed as part of a urinalysis.

- **urologist**
  a physician and surgeon who is specially trained in the diagnosis and treatment

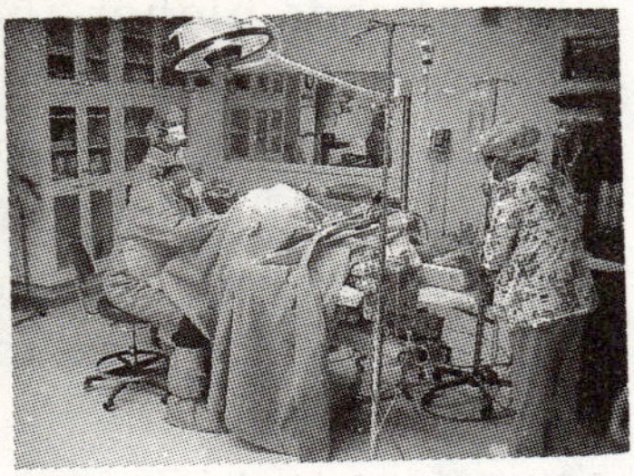

  of diseases of the male genital tract and urinary tract in patients of any age or sex.

- **urticaria**
  vascular reaction of the skin marked by the transient appearanceof smooth, slightly elevated patches (wheals, hives) that are redder or paler than the surrounding skin and often attended by severe itching.

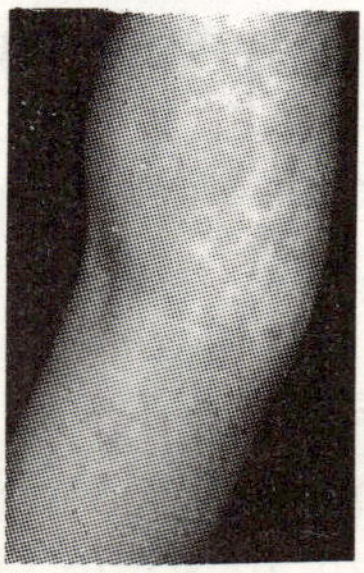

- **user fee**
  a fee which is collected only from those persons who use a particular service, as opposed to one collected from the public in general. User fees generally vary in proportion to the degree of use of the service.

- **USTs**
  underground storage tanks.

- **vacuole**
  membrane-bound cavity within a cell.

- **validity (of a measurement)**
  expression of the degree to which a measurement measures what it purports to measure. See **concurrent validity, construct validity, content validity, criterion validity, predictive validity.**

- **validity of a study**
  degree to which the inferences drawn from a study, especially generalisations extending beyond the study sample, are warranted when account is taken of the study methods, the representativeness of the study sample, and the nature of the population from which it is drawn. See **external validity, internal validity**.

- **vapour pressure**
  partial pressure of a liquid's vapour required to maintain the vapour in equilibrium with the condensed liquid or solid. Also referred to as saturation vapour pressure.

- **vaporisation**
  the change of a substance from a liquid to a gas.

- **vapour**
  the gas given off by substances that are solids or liquids at ordinary atmospheric pressure and temperatures.

- **vapour dispersion**
  the movement of vapour clouds or plumes in the air due to wind, gravity, spreading, and mixing.

- **vapour pressure**
  the pressure exerted by a saturated vapour above its own liquid in a closed container. Vapour pressures reported in millimetres of mercury (mm Hg) at 68 degrees F (20 degrees C), unless stated otherwise. Three facts are important to remember: (1) vapour pressure of a substance at 100 degrees F (38 degrees C) is always higher than the vapour pressure of the substance at 68 degrees F (20 degrees C). (2) Vapour pressures reported in mm Hg are usually very low pressures; 760 mm Hg is equivalent to 14.7 pounds per square inch. (3) the lower the boiling point of a substance, the higher its vapour pressure. Vapour pressures are useful (with evaporation rates) in learning how quickly a material becomes airborne within the workplace and thus how quickly a worker is exposed to it.

- **vapour recovery system**
  a system by which the volatile gases from gasoline are captured instead of being released into the atmosphere. Recovery systems may be required for gasoline stations in some cities and

other non-attainment areas.

- **variable costs**
  input costs that change as the nature of the production activity of its circumstances change, for example, as production levels vary.

- **vasoconstriction**
  decrease of the calibre of the blood vessels leading to a decreased blood flow. See **vasodilation**.

- **vasodilation**
  increase in the calibre of the blood vessels, leading to an increased blood flow. See **vasoconstriction**.

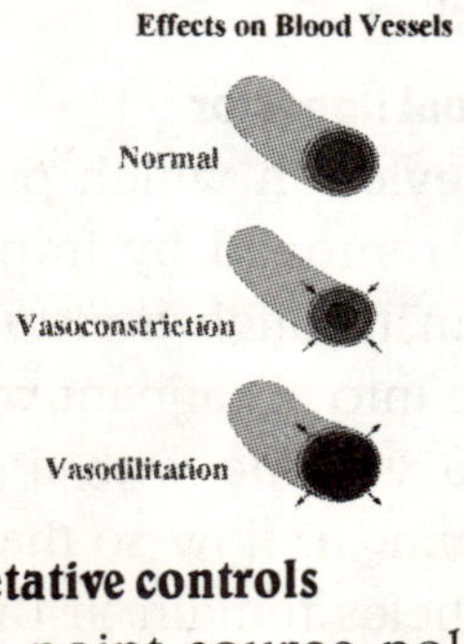

- **vegetative controls**
  non-point source pollution control practices that involve plants (vegetative cover) to reduce erosion and minimise the loss of pollutants.

- **vehicle**
  substance(s) used to formulate active ingredients for administration or use (general term for solvents, suspending agents, etc.). See **excipient**.

- **vena contracta**
  flow contraction with flow separation from the wall, usually occurring after constriction of a flow channel or just downstream of the entry point of an inlet.

- **venom**
  animal toxin generally used for self-defence or predation and usually delivered by a bite or sting. See **toxin**.

- **ventilation**
  1. process of supplying a building or room with fresh air.
  2. process of exchange of air between the ambient atmosphere and the lungs.
  3. in physiology, the amount of air inhaled per day.

4. oxygenation of blood.

- **ventricular fibrillation**
irregular heartbeat characterized by uncoordinated contractions of the ventricle.
- **venturi meter**
a device used to measure flow rate in a duct by measuring the pressure drop across a calibrated streamlined constriction.
- **vermicide**
substance intended to kill worms.
- **vermifuge**
substance that causes the expulsion of intestinal worms.
- **vertical elutriator**
a vertical channel that gravitationally retains or removes particles above a given size or size range and emits the remaining airborne particles.
- **vertigo**
dizziness, an illusion of movement as if the external world were revolving around an individual or as if the individual were revolving in space.
- **vesicant**
1. producing blisters on the skin.
2. substance that causes blisters on the skin.
- **vesicle**
1. small sac or bladder containing fluid.

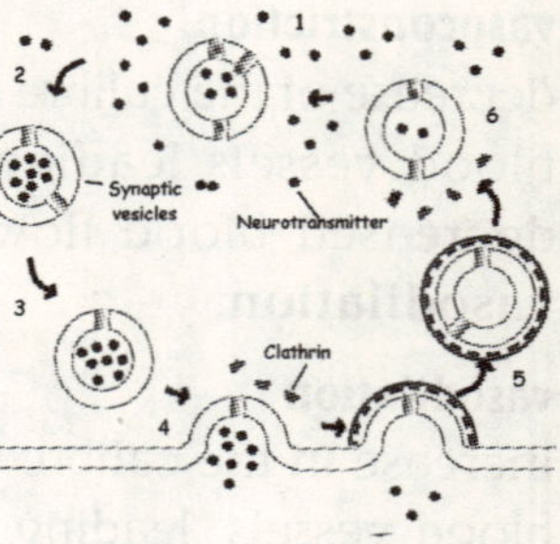

2. blisterlike elevation on the skin containing serous fluid.
- **virtual impactor**
a device in which particles are removed by impacting them through a virtual surface into a stagnant volume, or a volume with a slowly moving airflow, so that large particles remain in this volume while smaller particles are deflected with the bulk of the original air flow. The dichotomous impactor is a frequently used virtual impactor.

- **virus**
one of a group of minute infectious agents (much smaller than bacteria) lacking independent metabolism. It is able to replicate only within living host cells. A virus consists of a core of

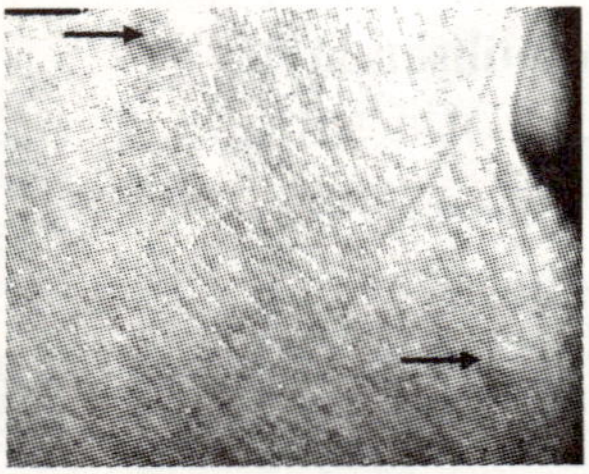

nucleic acid surrounded by an inner protein mantle, sometimes encircled by an outer protein envelope. Viral infections usually lead to the production of antibodies.

- **vital capacity**
maximum volume of gas that can be exhaled from the lung after maximum inhalation.

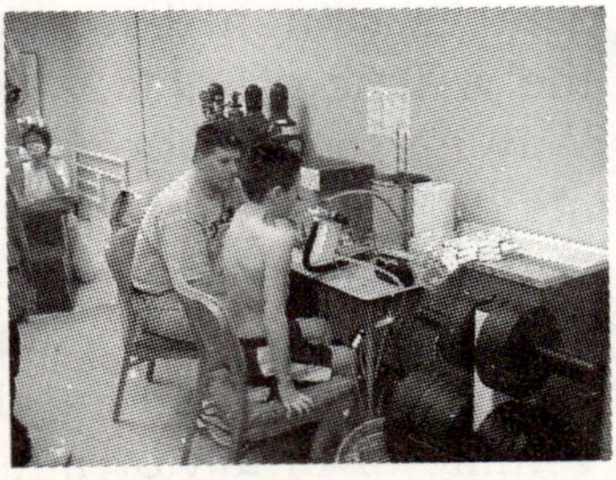

- **volatile**
evaporating readily at normal temperatures and pressures. The air concentration of a highly volatile chemical can increase quickly in a closed room.

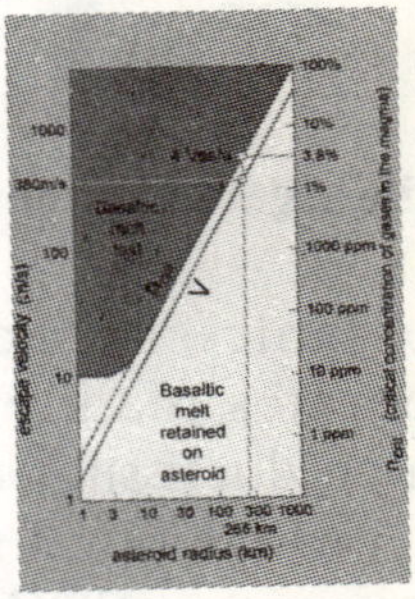

- **volatile acids**
acids produced during digestion. Fatty acids which are soluble in water and can be steam-distilled at atmospheric pressure. Also called 'organic acids'. Volatile acids are commonly reported as equivalent to acetic acid.

- **volatile liquids**
liquids which easily vaporize or evaporate at room temperatures.

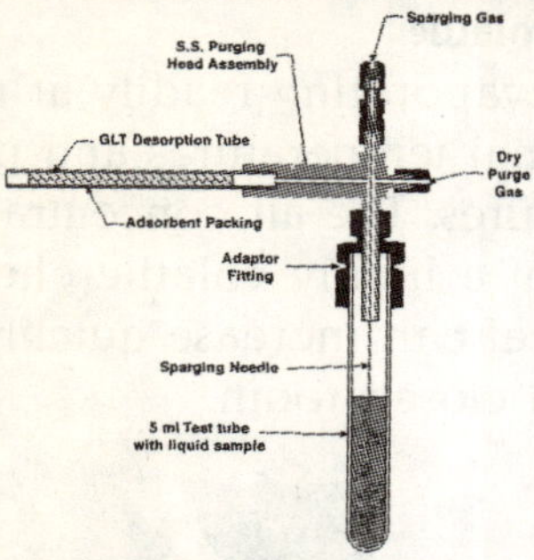

- **volatile organic compound (VOC)**

  an organic chemical that evaporates readily. Petroleum products such as kerosene, gasoline and mineral spirits contain VOCs. Chlorinated solvents such as those used by dry cleaners or contained in paint strippers are also VOCs. See '**organic**' and '**volatile**'.

- **volatile solids**

  those solids in water or other liquids that are lost on ignition of the dry solids at 550 degrees C.

- **volatile substances**

  material that is readily vaporisable at a relatively low temperature.

- **volatile synthetic organic chemicals**

  chemicals that tend to volatile or evaporate from water.

- **volatile.**

  readily vaporisable at a relatively low temperature.

- **volatility**

  measure of a material's tendency to vaporise or evaporate at ambient routine conditions.

- **volatilisation**

  loss of a substance through evaporation.

- **voltage**

  the electrical pressure available to cause a flow of current (amperage) when an electrical circuit is closed.

- **volume of distribution**

  apparent (hypothetical) volume of fluid required to contain the total amount of a substance in the body at the same concentration as that present in the plasma assuming equilibrium has been attained. See **plasma**.

- **volumetric**

  a measurement based on the volume of some factor. Volumetric titration is a means of measuring unknown concentrations of water quality indicators in a sample by determining the volume of titrant or liquid reagent needed to complete particular reactions.

- **vortex**
  a revolving mass of water which forms a whirlpool. This whirlpool is caused by

water flowing out of a small opening in the bottom of a basin or reservoir. A funnel-shaped opening is created downward from the water surface.

- **wall loss**
  deposition of particles in a sampler on surfaces other than those designed for particle collection.

- **waste**
  anything that is discarded deliberately or otherwise disposed of on the assumption that it is of no further use to the primary user.

- **waste heat**
  heat that escapes to the atmosphere during combustion processes. Minimising and recapturing waste heat is a valuable energy conserving strategy. For example, in homes there is heat in grey water, flues from wood-burning stoves, and heat pump fluids that can be captured for other uses such as preheating domestic hot water.

- **waste load allocation (WLA)**
  the portion of a receiving water's total maximum daily load that is allocated to one of its existing or future point sources of pollution

- **waste water**
  water for which disposal is more economical than use at the time and point of its occurrence. Waste water to one user may be a desirable supply to the same or another user at a different location. It may be waste water because of its quality, quantity, or the time of its occurrence.

- **wastewater treatment plant**
  a facility containing a series of tanks, screens, filters, and other processes by which pollutants are removed from water. Most treatments include chlorination to at-

tain safe drinking water standards.

• **wasting syndrome**
disease marked by weight loss and atrophy of muscular and other connective tissues that is not directly related to a decrease in food and water consumption.

• **water budget**
a summation of inputs, outputs, and net changes to a particular water resource system over a fixed period. (Also, water balance model).

• **water effects ratio (WER)**
a correction factor used to adjust a state or regional water quality standard to account for differences in the toxicity of a specific pollutant between laboratory water and site water.

• **water hammer**
the sound like someone hammering on a pipe that occurs when a valve is opened or closed very rapidly. When a valve position is changed quickly, the water pressure in a pipe will increase and decrease back and forth very quickly. This rise and fall in pressures can do serious damage to the system.

• **water pollution**
industrial and institutional wastes, and other harmful or objectionable material in sufficient quantities to result in a measurable degradation of the water quality.

• **water purveyor**
an agency or person that supplies water (usually potable water).

• **water quality**
a term used to describe the chemical, physical, and biological characteristics of water with respect to its suitability for a particular use.

- **water quality standard (WQS)**
  the combination of a designated use and the maximum concentration of a pollutant which will protect that use for any given body of water. For example, in a trout stream, the concentration of iron should not exceed 1 mg/l.

- **water solubility**
  the maximum possible concentration of a chemical compound dissolved in water. If a substance is water soluble it can very readily disperse through the environment.

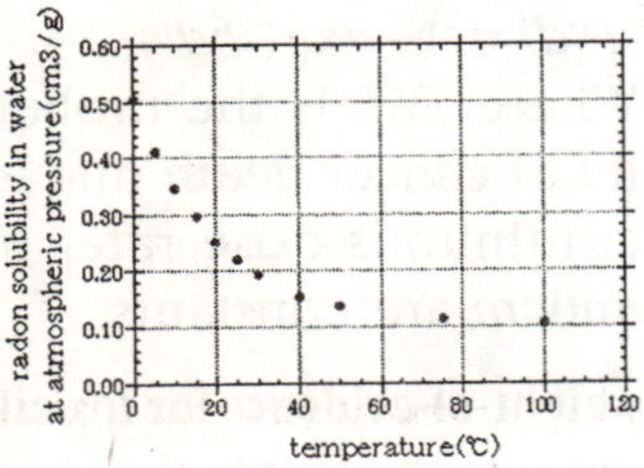

- **water storage pond**
  an impound for liquid wastes, so designed as to accomplish some degree of biochemical treatment of the wastes.

- **water supplier**
  one who owns or operates a public water system.

- **water supply**
  any quantity of available water.

- **water supply system**
  the system for the collection, storage, treatment and distribution of potable or other kinds of water from the sources of supply to the consumer.

- **water table**
  the boundary between the saturated and unsaturated zones. Generally, the level to which water will rise in a well (except artesian wells).

- **water treatment lagoon**
  an impound for liquid wastes, so designed as to accomplish some degree of biochemical treatment of the wastes.

• **water well**
an excavation where the intended use is for the location, acquisition, development, or artificial recharge

of groundwater (excluding sandpoint wells).

• **waterborne disease outbreak.**
the significant occurrence of acute infectious illness, epidemiologically associated with the ingestion of water from a public water system that is deficient in treatment, as determined by the appropriate local or state agency

• **watershed**
the land area that drains into a stream. An area of land that contributes runoff to one specific delivery point. Large watersheds may be composed of several smaller 'subsheds', each of which contributes runoff to different locations that ultimately combine at a common delivery point.

• **watt**
a unit of power equal to one joule per second. The power of a current of one ampere flowing across a potential difference of one volt.

• **weibull model**
dose-response model of the form:

$$P(d) = 1 - \exp(-bd^m)$$

Where $P(d)$ is the probability of cancer death due to a continuous dose rate, d, b and m are constants.

• **weight-of-evidence for toxicity**
extent to which the available biomedical data support the hypothesis that a substance causes a defined toxic effect such as cancer in humans.

• **weir**
1. a wall or plate placed in an open channel and used to measure the flow of wa-

ter. The depth of the flow over the weir can be used to calculate the flow rate, or a chart or conversion table may be used.

2. a wall or obstruction used to control flow (from settling tanks and clarifiers) to assure uniform flow rate and avoid short-circuiting.

- **weir diameter**

many circular clarifiers have a circular weir within the outside edge of the clarifier. All the water leaving the clarifier flows over this weir. The diameter of the weir is the length of a line from one edge of a weir to the opposite edge and passing through the centre of the circle formed by the weir

- **weir loading**

a guideline used to determine the length of weir needed on settling tanks and clarifiers in treatment plants. Used by operators to determine if weirs are hydraulically (flow) overloaded. Weir Loading (GPM/ft) = Flow (GPM)/ Length of Weir (ft)

- **well**

a bored, drilled, or driven shaft, or a dug hole, whose depth is greater than the largest surface dimension and whose purpose is to reach underground water supplies or oil, or to store or bury fluids below ground.

- **well field**

area containing one or more wells that produces usable amount of water.

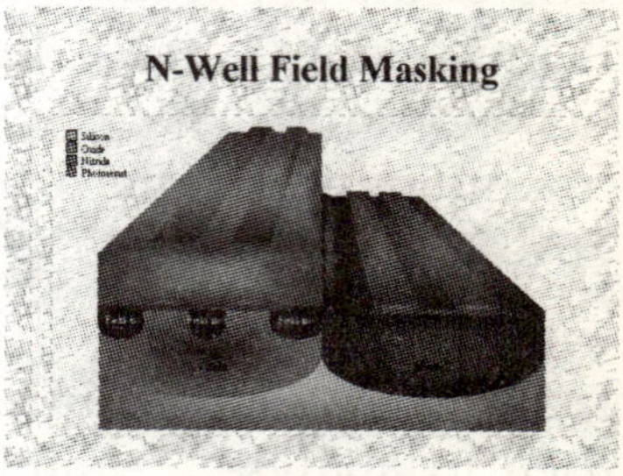

- **well monitoring**

the measurement, by on-site instruments or laboratory methods, of the quality of water in a well.

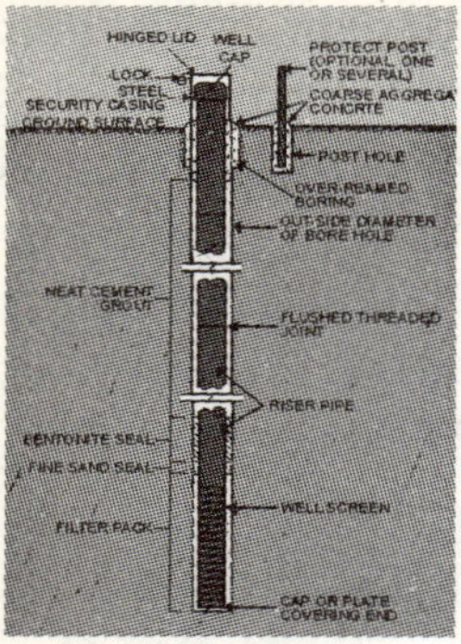

- **well plug**

  a watertight and gastight seal installed in a bore hole or well to prevent movement of fluids.

- **western blot**

  a method of immobilising DNA (viral) fragments on a support medium where they can be identified using specific antibodies. One of the applications of the procedure is the determination of the specificity of antibodies to the AIDS virus.

- **wet chemistry**

  laboratory procedures used to analyse a sample of water using liquid chemical solutions (wet) instead of, or in addition to, laboratory instruments.

- **wetlands**

  areas that are soaked or flooded by surface or ground water frequently enough or for sufficient duration to support plants, birds, animals, and aquatic life. Wetlands generally include swamps, marshes, bogs, estuaries, and other inland and coastal areas, and are federally protected. Wetlands frequently serve as recharge/discharge areas and are known as 'nature's kidneys' since they help purify water. Wetlands also have been referred to as natural sponges that absorb flood waters, functioning like natural tubs to collect overflow. Wetlands are important wildlife habitats, breeding grounds, and nurseries because of their biodiversity. Many endangered species as well as countless estuarine and marine fish and shellfish, mammals, waterfowl, and other migratory birds use wetland habitat for growth, reproduction, food, and shelter. Wetlands are among the most fertile, natural ecosystems in the world since they produce great volumes of food (plant material).

- **whole effluent toxicity (WET)**
the total toxic effect of an effluent measured directly with a toxicity test .

- **whole sediment**
sediment and associated pore water that have had minimal manipulation.

- **wildlife refuge**
an area designated for the protection of wild animals, within which hunting and fishing are either prohibited or strictly controlled.

- **wire-to-water efficiency**
the efficiency of a pump and motor together. Also called the overall efficiency.

- **withdrawal**
the process of taking water from a source and conveying it to a place for a particular type of use.

- **withdrawal effect**
adverse event following withdrawal from a person or animal of a drug to which they have been chronically exposed or on which they have become dependent.

- **working zone**
space measuring up to 2 m over the level of the floor or platform that contains a worker's permanent or temporary station.

- **x-disease**
hyperkeratotic disease in cattle following exposure to chlorinated dibenzo-*p*-dioxins, naphthalenes and related compounds.

- **xenobiotic**
1. strictly, any substance interacting with an organism that is not a natural component of that organism. See **exogenous substance**, **foreign substance or compound**.
2. man-made compounds with chemical structures foreign to a given organism.

- **yield**
the quantity of water (expressed as a rate of flow-GPM, GPH, GPD, or total quantity per year) that can

be collected for a given use from surface or groundwater sources. The yield may vary with the use proposed, with the plan of development, and also with economic considerations. Also see **safe yield.**

- **zeta potential**
  in coagulation and flocculation procedures, the difference in the electrical charge between the dense layer of ions surrounding the particle and the charge of the bulk of the suspended fluid surrounding this particle. The zeta potential is usually measured in millivolts.

- **zone of aeration**
  the comparatively dry soil or rock located between the ground surface and the top of the water table.

- **zone of saturation**
  the soil or rock located below the top of the groundwater table. By definition, the zone of saturation is saturated with water. Also see **water table.**

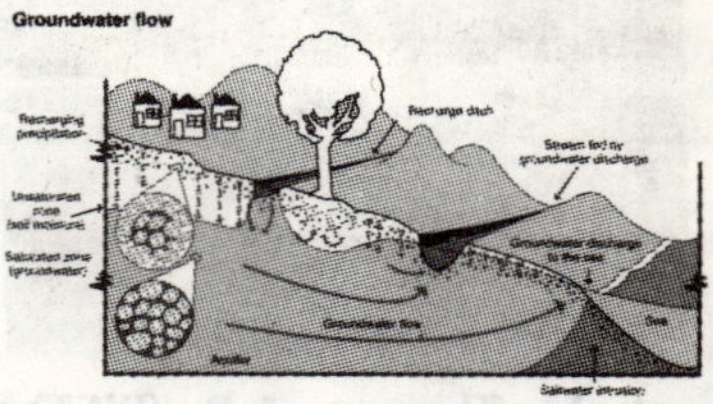

- **zoocide**
  substance intended to kill animals.

- **zooplankton**
  small, usually microscopic animals (such as protozoans), found in lakes and reservoirs.

- **zygote**
  1. cell such as a fertilised egg resulting from the fusion of two gametes.

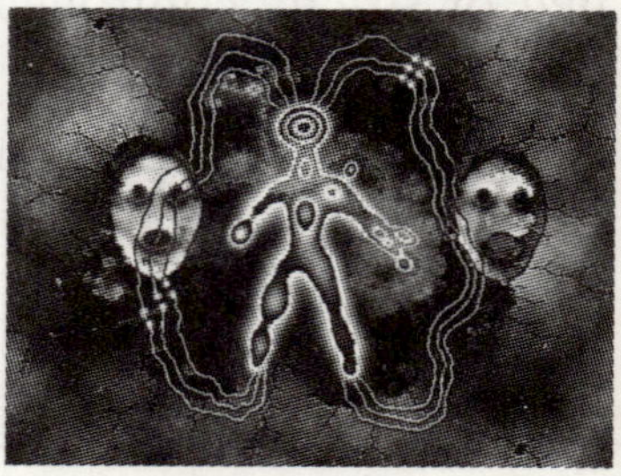

  2. cell obtained as a result of complete or partial fusion of cells produced by meiosis.

Notes

Notes

Notes

Notes

Notes

Notes